Praise for Soul Made Flesh

"Carl Zimmer's illuminating book charts a fascinating chapter in the soul's journey. . . . Zimmer successfully communicates his enthusiasm for the energetic minds and busy pens of his heroes. His book is timely."

—*The New York Times Book Review*

"Ravaged by religious wars and capricious monarchs, 17th-century England was a kingdom in chaos. Against this bloody backdrop, Zimmer recounts physician Thomas Willis' momentous discovery that the brain—previously dismissed as 'a bowl of curds'—is the seat of human consciousness and memory. This page-turner is a tribute to the heretical thinkers who decoded nature by relying on direct observation rather than received opinion."

—*Wired* magazine

"An uncommonly literate look at a little-explored side of scientific history, and a thumping good read at that."

—Timothy Ferris, author of *The Whole Shebang* and *Coming of Age in the Milky Way*

"We live in what Carl Zimmer, one of our most gifted science writers, calls a Neurocentric Age . . . Zimmer describes . . . a kind of second Copernican revolution—one inside the body. . . . Thrilling . . . Zimmer's nimble survey of the intellectual landscape of the 17th century [is] a top-notch work of popular science, chock-full of fascinating lore and inspired quotations. . . . Hosts of knotty concepts are treated to lucid descriptions, and his fluent prose and vivid narration prove themselves as much at home among the complex historical and political crosscurrents of the 17th century as they are with finely tuned accounts of biochemistry or MRI scanners."

—Ross King, author of *Brunelleschi's Dome*, in the *Los Angeles Times*

"In *Soul Made Flesh,* Carl Zimmer gives a remarkable, beautiful account of England's 'genius century.' Zimmer brings Willis to life—his prose, as always, is clear, vivid, and arresting—and reminds us how startling and revolutionary his discoveries were."

—Oliver Sacks

"A deep and contextualized exploration of two millennia's worth of human theories about consciousness and the soul. . . . [Zimmer's] wide-ranging narrative reaches from the days of Aristotle to a 21st-century lab in the basement of a Princeton University building. The central figure in Zimmer's tale is the oft-overlooked 17th-century scientist Thomas Willis, a figure of fascinating contradictions. . . . In the end, however, this book is less about Willis in particular than about the evolving metaphysics of the soul in general, and the reader is left with a better picture of the roots of the modern understanding of the self as well as a familiarity with one of the unsung heroes of the scientific revolution."

—*Publishers Weekly*

"A gifted science writer, Zimmer recounts Willis' singular achievement in a narrative that illuminates not only the scientific revolution in medicine but also the cross-grained personality of one of the chief revolutionaries. . . . A remarkable fusion of scientific history and cultural analysis."

—*Booklist* (starred review)

"*Soul Made Flesh* tells the fascinating story of how people first became aware of one of the most radical thoughts the human mind has ever had to think. The writing is vivid and literate, the story compelling, and the modern implications drawn out with skill and verve."

—Steven Pinker, bestselling author of *How the Mind Works* and *The Blank Slate*

"Carl Zimmer clarifies and illuminates the story of a fascinating thinker. By focusing on a single player in the vast spectacle that was the Scientific Revolution, and telling his story so well, Zimmer gives us insights into the age when Alchemy gave way to modern science. But this is not only a history book, for readers with an interest in consciousness and the brain will find much here that applies to research going on today."

—Neal Stephenson, author of *Snow Crash* and *Cryptonomicon*

"Instructive and engaging. . . . Like *The Lunar Men*, Jenny Uglow's recent tour de force history of 18th century British scientists, Zimmer's book is a study in intellectual comradeship and cooperation, and how thinkers are shaped by their milieu."

—*Newsday*

"Few writers can bring back the odor and the sense of time that are present during historic discoveries. Few can capture the extent of human ignorance that is present and is about to be illuminated. Carl Zimmer writes with a rare, captivating skill that brings one back to that place. This is a must read."

—Michael S. Gazzaniga, Ph.D., David T. McLaughlin
Distinguished Professor in Cognitive Neuroscience,
Dartmouth College, and author of *Nature's Mind*

"*Soul Made Flesh* provides an account of the first big steps toward an understanding of how the brain makes mind. . . . A fine intellectual history . . . full of drama."

—*Natural History*

"Wry and engaging . . . Zimmer plunges us elbow deep into the messy realities of 17th-century medicine."

—*Hartford Courant*

"The main parallels that can be drawn between politics, religion, science, and human behavior then and now add unexpected dividends to this engaging narrative. Absorbing and thought-provoking."

—*Kirkus Reviews*

"Zimmer draws a vivid picture of the background against which Willis and other scientists of the time worked."

—*Scientific American*

"A panoramic history of England during a period of political upheaval, civil war, religious ferment, plague and assorted other stresses."

—*St. Petersburg Times*

"A fascinating look at the medical pioneer who dared to explore the seat of the soul. . . . Zimmer paints a vivid picture of the life and times of this stubborn 17th-century trailblazer. . . . Willis left behind a legacy more far-reaching than he could have dreamed. We are in his debt, and in Zimmer's as well for his hugely entertaining portrait of this scientific hero."

—*BookPage*

"Carl Zimmer brings to astonishingly vivid life a momentous turning point in science's history, when a band of brave British anatomists revealed that our memories, visions, fears, dreams—our very souls—spring from a three-pound lump of flesh in our skulls. One of our best science journalists turns out to be a skilled historian as well."

—John Horgan, author of *The End of Science* and *Rational Mysticism*

"Peppered with amusing anecdotes about the principal philosophers who struggled with the nature of the 'sensitive' and 'rational' soul."

—*Nature Neuroscience*

"A breath-taking journey. . . . Elegant and enthralling. . . . A luminous narrative of lively characters and of the brain's desire to know itself."

—*Billerica (MA) Minuteman*

"An award-winning science writer narrates the little-known story of Thomas Willis. He discusses the context of 17th-century views, politics, and the insights of other scientists and philosophers—all leading to a new scientific paradigm."

—*Book News, Inc.*

"*Soul Made Flesh* makes good reading."

—*New England Journal of Medicine*

"Witty and erudite . . . Carl Zimmer has faithfully recounted the long saga of the problem of explicating the brain and how it became more scientifically based as a result of the studies of Willis and his colleagues."

—*Brain: A Journal of Neurology*

"As intellectual history, [*Soul Made Flesh*] is a superb read for anyone and a must read for neuroscientists."

—*Science & Theology News*

"*Soul Made Flesh* belongs in all libraries and would make a superb gift for the neuroscientist or anyone who works with the brain. Zimmer's knowledge of the period and depiction of the individuals who were responsible for many important scientific theories make for an exciting read. This was a hard book to put down."

—*The Journal of Clinical Investigation*

"[An] engrossing account of how mechanism replaced spirit, and how the soul was finally banished to the stuff of dreams . . . Zimmer's book has many virtues [and] is adept at bringing together the luminaries and other actors of the time."

—Simon Conway Morris, *BioScience*

"A much-needed book on an extraordinary 17th-century figure, who did landmark work that deserves to be brought to everyone's attention."

—Lisa Jardine, author of *The Curious Life of Robert Hooke,* in *The Times* (London)

Other Books by Carl Zimmer

Evolution: The Triumph of an Idea

Parasite Rex: Inside the Bizarre World of Nature's Most Dangerous Creatures

At the Water's Edge: Fish with Fingers, Whales with Legs, and How Life Came Ashore but Then Went Back to Sea

Soul Made Flesh

The Discovery of the Brain—and How It Changed the World

CARL ZIMMER

FREE PRESS
New York London Toronto Sydney

*f*P
FREE PRESS
A Division of Simon & Schuster, Inc.
1230 Avenue of the Americas
New York, NY 10020

First Free Press trade paperback edition 2005

FREE PRESS and colophon are trademarks
of Simon & Schuster, Inc.

For information regarding special discounts for bulk purchases,
please contact Simon & Schuster Special Sales:
1-800-456-6798 or business@simonandschuster.com

Designed by Nancy Singer Olaguera

Manufactured in the United States of America

10 9 8 7 6 5 4 3 2 1

The Library of Congress has cataloged the hardcover edition as follows:
Zimmer, Carl.
Soul made flesh : the discovery of the brain—and how it changed the world / Carl Zimmer.
p. cm.
Includes bibliographic references and index.
1. Brain. 2. Brain—Miscellanea. 3. Brain—History. I. Title.

QP376.Z555 2004
612.8'1—dc22 2003063144

ISBN 0-7432-3038-8
0-7432-7205-6 (Pbk)

Illustrations: Pages 2, 8, 24, 42, 56, 116, 146, and 188, reproduced with the permission of William Feindel, editor of Thomas Willis, *The Anatomy of the Brain and Nerves* (1664), tercentenary ed., vol. 2 (Montreal: McGill University Press, 1965). Pages 82 and 168, British Library. Pages 208 and 236, Bodleian Library, Oxford University. Page 260, British Museum.

To Charlotte,
whose soul grew along with this book

To explicate the uses of the Brain seems as difficult a task as to paint the Soul, of which it is commonly said, that it understands all things but itself.

—Thomas Willis, *The Anatomy of the Brain and Nerves* (1664)

Contents

Soul Made Flesh

Introduction

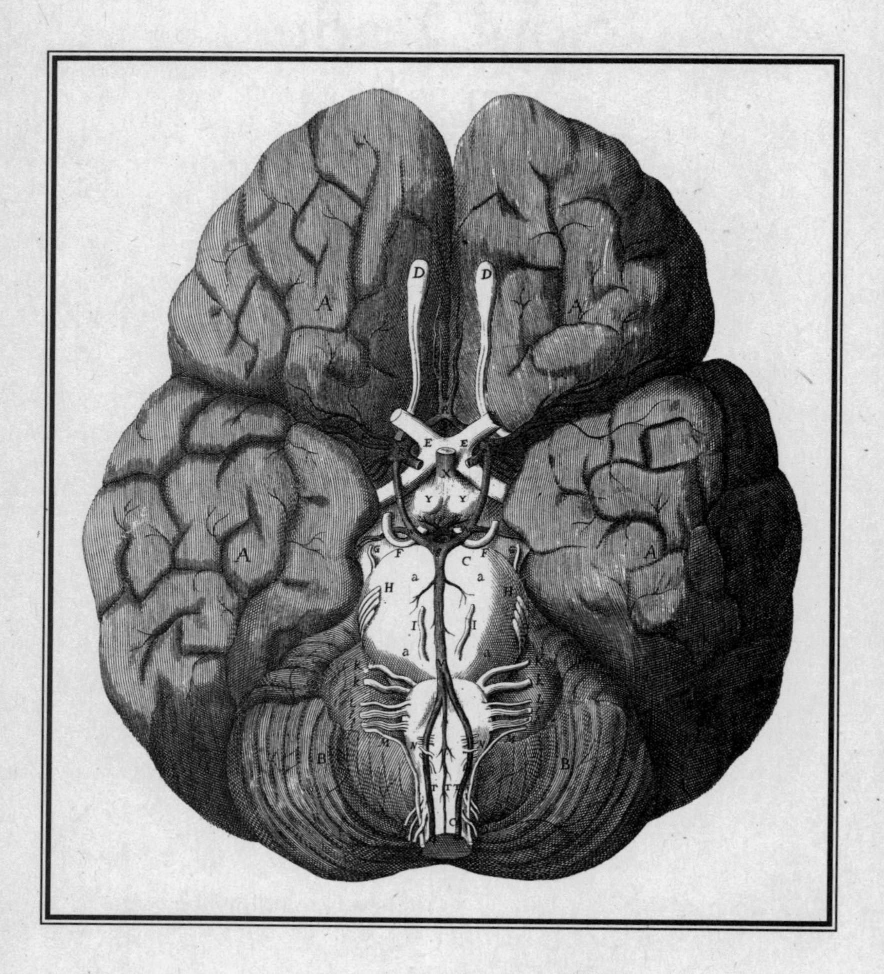

A FULL HUMAN BRAIN, DRAWN BY CHRISTOPHER WREN AND INCLUDED IN *THE ANATOMY OF THE BRAIN AND NERVES.*

A Bowl of Curds

To imagine a time and place—say, the city of Oxford on a summer day in 1662—you have to engage not only the mind's eye and ear but also the mind's nose. The warm odor of malt and corn flour rises from the boats landing at the wharves along the Thames. The stink of cured fish hanging in fishmongers' stalls mixes with the soft smell of bread in the bakeries. The smell of manure is everywhere, in the open sewers, on the town common where cows graze, in the streets where horses haul wagons and coaches. Sometimes a coach rolls through the narrow gate of one of Oxford's colleges, to be swallowed up behind a high, windowless stone wall. The chimneys of the college kitchens relay smoke signals to the surrounding neighborhoods, carrying the smell of roasting capon and mutton or perhaps a goose stolen from a nearby village by students.

On a summer day the perfume of the surrounding fens and meadows drifts into the city and mixes with the exotic scents of the physik garden on the High

Street, a home to exotic species such as leopard's bane, mimosa trees, Virginian spiderwort, and scorpion grass. Botanists gather their leaves and seeds and roots and carry them to an apothecary's shop to be ground down, cooked, distilled, and mixed with sharp-odored hartshorn or spirits of wine.

Every building in Oxford has an internal signature of smells: the incense burning in the churches once again, now that the Puritans have been routed and the monarchy restored; the roasted beans in the new coffeehouse on High Street; the foul reek of the prisons, where thieves, Quakers, and various enemies of King Charles II languish together. But the strangest smells in all of Oxford can be found off the main thoroughfares, on Merton Street. Across the street from the gates of Merton College is a medieval two-story house known as Beam Hall. Its odors are almost unbearable: a reeking blend of turpentine and the warm, decaying flesh of dissected dogs and sheep, along with an aroma that none but a handful of people in Oxford—in the world, even—would recognize as that of a nobleman's decapitated and freshly cracked open head.

The room where his body is being dissected is something between a laboratory and a butcher's shop. Knives, saws, and gimlets hang on the walls, along with pliers and razors, brass and silver probes, pincers, bugles for inflating membranous sacs, curved needles, augurs, mallets, wimbles, and bodkins. Syringes and empty quills sit on a table, along with bottles of tincture of saffron and a simple microscope, illuminated by an oil lamp and a globe of brine. Hearts rest at the bottom of jars, pickled. On a long table lies the corpse, surrounded by a crowd of natural philosophers. Depending on the day, the audience may include a mathematician who is laying the groundwork for calculus or a chemist who is in the process of turning alchemy into a modern science. Astronomers, doctors, and ministers come to watch. They all stare intensely, because they know they are part of an unprecedented experience. They are anatomizing the soul.

An inner circle of men stands closest to the body. Christopher Wren, thirty years old and not yet England's great architect, studies the exposed flanges and curves of the skull. He can sketch bowels and hearts as beautifully as he will later sketch a cathedral dome. Richard Lower, who in a few years will perform the first successful blood transfusion in history, severs the nobleman's carotid arteries and slices the gristly cartilage between his cervical vertebrae. The finest dissector in all Europe, he serves as assistant to another man in the inner circle, the owner of Beam Hall, the man who has assembled this herd of natural philosophers within its walls—a short, stammering physician with hair that one neighbor describes in his diary as being "like a dark red pigge." His name is Thomas Willis.

Willis has brought these men together this day in 1662 in order to come to a new understanding of the brain and nerves. He and Lower strip the skin and then cut away the inner mask of muscle. They saw off the bones of the skull, prying away each one with a penknife or a pair of scissors. They snip the nerves that tether the brain to the eye and nose. All that is left is the brain encased in its membranes. Next Willis and Lower turn the brain upside down and gently peel away the membranes so as not to damage the delicate nerves and blood vessels at its base. Furrowed and lobed, the brain is liberated, and Willis holds it aloft for his audience to see.

Today, when we look at a brain, we see an intricate network of billions of neurons in constant, crackling communication, a chemical labyrinth that senses the world outside and within, produces love and sorrow, keeps our hearts beating and lungs breathing, composes our thoughts, and constructs our consciousness. To most people in 1662, however, this would all have sounded quite absurd. When the contemporary English philosopher Henry More wrote about the brain, he declared that "this lax pith or marrow in man's head shows no more capacity for thought than a cake of suet or a bowl of curds." The brain, More wrote, was a watery, structureless substance which could not contain the complex workings of the soul. The idea that the frail flesh in our heads was capable of the soul's work was more than just absurd. It bordered on atheism. If reason, devotion, and love were the work of mortal flesh instead of immaterial spirit, then what would become of the soul after death? What need was there for a soul at all? Henry More put the matter simply: "No spirit, no God."

Exactly what spirits and soul consisted of and where they could be found were questions that had been asked and re-asked for well over two thousand years. At the beginning of the seventeenth century, most Europeans would have agreed that the soul was the immortal, immaterial essence of a person, which would be saved or damned by God. But the same word could also refer to an intelligence at work throughout the entire body—making it grow to its destined shape, making it warm and alive, reproducing its form in children. Spirits were the instruments used by the soul and body to reach their goals. For many philosophers, alchemists, apothecaries, and mystics, the cosmos also had a soul, which channeled spirits through planets and stars to enact its will—spirits that could be harnessed by magic or alchemy. With each breath, the world's spirits entered the human body and infused it with life and intelligence, uniting the soul of the microcosm with the soul of the macrocosm.

As widely held as all these beliefs were in 1600, they were being steadily undermined. By the end of the seventeenth century, they would all be either obliterated or fatally wounded, and Thomas Willis and his friends were playing a crucial role in the transformation. Their grisly work in Beam Hall was the first modern investigation of the nervous system. Whenever Willis held a brain in his hands and described it to his audience, he did not limit himself to the branchings of nerves and other anatomical details. He showed how the brain's intricate structures could form memories, hatch imaginations, experience dreams. He reconceived thoughts and passions as a chemical storm of atoms. Willis called his brain project a "doctrine of the nerves" and coined a new Latin word for it: *neurologie.*

Although Willis and his friends were establishing the modern science of the brain, they do not fit the modern definition of a scientist. Some were alchemists who searched out the philosopher's stone so as to be able to communicate with angels. Some were physicians who recommended carved-up puppies for clearing the skin. All of them were seeking signs of God's work in a universe that had become terrifying and alien. They were scarred by civil war and hoped that a new conception of the brain would bring order and tranquility to the world. Their claims were often accepted not so much because they were true (which, fairly often, they were not), but because the world itself had developed an appetite for them.

These men of Oxford ushered in a new age, one in which we still live—call it the Neurocentric Age—in which the brain is central not only to the body but to our conception of ourselves. The seventeenth century saw many scientific revolutions, but in some ways the revolution of the brain is its most shattering triumph—and its most intimate. It created a new way of thinking about thinking and a new way of conceiving the soul. Today, some three hundred forty years later, the Neurocentric Age is more deeply entrenched than ever. At the beginning of the twenty-first century, thousands of neuroscientists follow Willis's trail. They continue to dismantle the brain, but they don't have to pull it from a corpse to do so. Instead, they can scan the positronic glow of neurons recalling the faces of friends, searching for a word, generating anger or bliss, or reading the minds of others. These scientists have started to isolate the molecules that these neurons trade and are manipulating them with drugs.

To some extent, we have become comfortable with this new brain. Few will deny that the workings of our minds are the product of billions of neurons organized into clusters and networks, trading trillions of signals with one another every second. We demonstrate our comfort by buying billions of dollars of drugs in the hope of lifting our mood, calming our jitters, or otherwise modifying who we are, simply by boosting or squelching the right neurochemical signals.

This comfort may have come too easily. The big business of brain drugs belies science's enormous ignorance about the organ. The maps that neuroscientists make today are like the early charts of the New World with grotesque coastlines and blank interiors. And what little we do know about how the brain works raises disturbing questions about the nature of our selves. In many ways, we are still standing in the circle at Beam Hall, with the odor of discovery in our noses, looking at the brain and wondering what this strange new thing is that Thomas Willis has found.

Chapter One

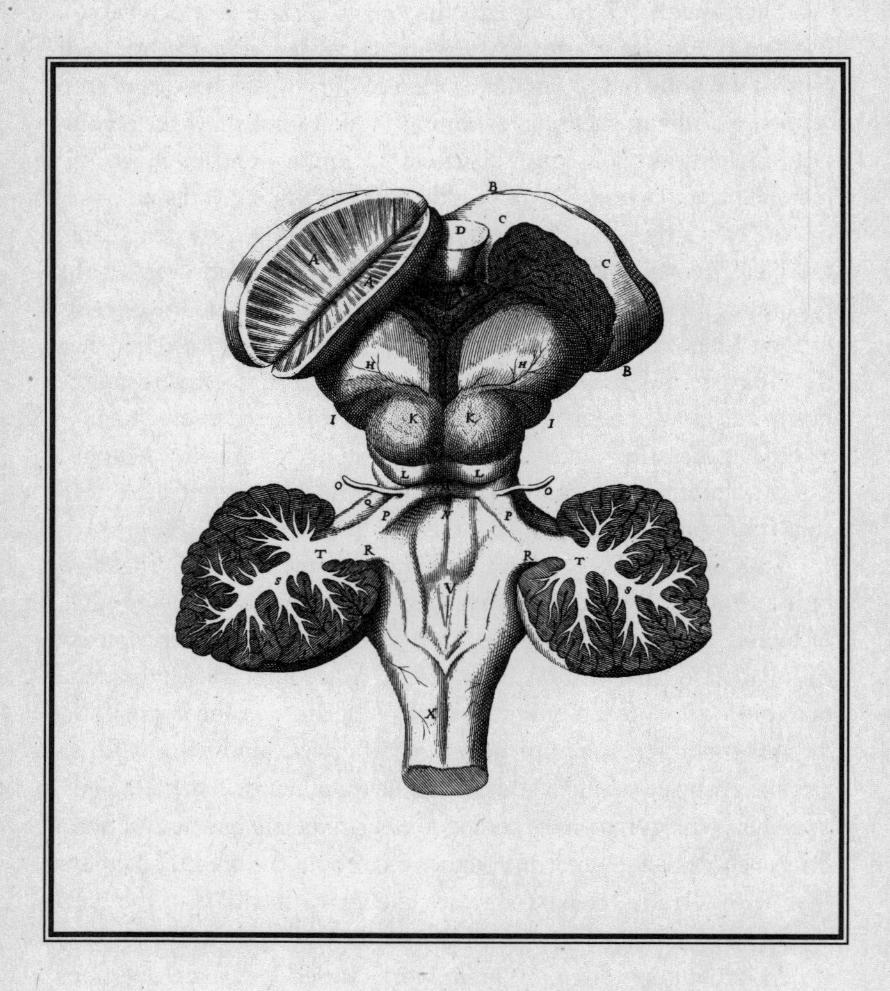

A SHEEP'S BRAIN STEM, FROM *THE ANATOMY OF THE BRAIN AND NERVES.*

Hearts and Minds, Livers and Stomachs

Thomas Willis was not the first person to take the brain out of its skull. The oldest records of the procedure come from ancient Egypt, four thousand years ago. The Egyptian priests who performed it did not hold up the brain and praise its power, however. Instead, they snaked a hook up the nose of the cadaver, broke through the eggshell-thin ethmoid bone, fished out the brain shred by shred until the skull was empty, and then packed the empty space with cloth.

The priests disposed of the brain while preparing the dead for the journey into the afterlife. The heart, by contrast, stayed in the body, because it was considered the center of the person's being and intelligence. Without it, no one could enter the afterlife. The jackal-headed god, Anubis, would place the deceased's heart in a scale, balancing it against a feather. The ibis-headed god, Thoth, would then ask the heart forty questions about the life of its owner. If the heart proved to be heavy with guilt, the deceased would be fed to the Devourer. If the heart was free of sin, the deceased would go to heaven.

It is difficult today to understand how the brain could be so dismissed, but throughout ancient times many people thought it unimportant. Others prized the brain but saw it not as a network of cells that produces language, consciousness, and emotions. They saw it as a shell of pulsing phlegm encasing empty chambers which whistled with the movement of spirits passing through. These two conceptions were powerful enough to guide Western thinking for thousands of years.

Some of the earliest philosophers of ancient Greece followed the Egyptian tradition. Empedocles described the soul as the thing that thinks, feels pleasure and pain, and gives the living body its warmth. At death, it leaves the body and searches for another home in a fish or a bird or even a bush; during its time in the human body, it resides around the heart.

But around 500 B.C., the Greek philosopher Alcmaeon lifted his gaze from the heart to the head, declaring that "all the senses are connected to the brain." Those words were a milestone in the history of science, but twenty-five hundred years later it's easy to misinterpret them. To begin with, Alcmaeon and his followers didn't even know that nerves existed. Few physicians had even seen these pale threads running through the body, because Greeks in general were reluctant to cut open cadavers. They were too worried that the souls of the dissected would not find rest in the afterlife until they got a proper burial. Alcmaeon reportedly cut the eye out of a dead animal's head and saw channels penetrating the skull. Like other ancient Greeks, he probably pictured channels in the body filled with spirits (or *pneumata*). These spirits were made of air, which was one of the four elements of the cosmos, along with fire, earth, and water. Each time a person took in a breath, these spirits were believed to flow into the nose, through the recesses of the brain, and into the body.

Alcmaeon's ideas helped shape early Greek medicine. In addition to spirits, physicians also came to believe that the body was composed of combinations of the elements known as humors. These four fluids—yellow bile, black bile, blood, and phlegm—each had its own qualities of moistness, dryness, heat, cold, and so on. The physician Hippocrates taught that good health was a matter of balancing the humors. If the brain, which was made of moist phlegm, became too moist, epilepsy might follow. If the phlegm moved from the brain to other parts of the body, tuberculosis or other diseases might strike.

Alcmaeon attracted a following not only among physicians but among philosophers as well. The most important of these was Plato, who gave the brain a central place in the cosmos. In his dialogue *Timaeus,* Plato described the cosmos as a living thing created by a divine craftsman, complete with its own immortal soul. The divine craftsman gave lesser gods the task of creating human beings, which they designed as the cosmos in miniature: with an immortal soul cloaked in a mortal body welded from the four elements. The gods began their work by creating the head, which they made spherical, like the cosmos. The divine seed was planted in the brain, where it could sense the world through the eyes and ears and then reason about it. This reasoning was the divine mission of the human soul; it would be able to reproduce the harmony and beauty of the cosmos in its own thoughts.

Into the rest of the body the gods inserted souls "of another nature," as Plato called them. In the guts dwelled "the part of the soul which desires meats and drinks and the other things of which it has need by reason of the bodily nature." This so-called vegetative soul was responsible for the body's growth and nutrition and also for its lower passions—its lusts, desires, and greed. To cage this wild beast, the gods built a wall—the diaphragm—separating it from a superior soul, which Plato located in the heart. The vital soul "is endowed with courage and passion and loves contention," Plato wrote. Along with the blood, the vital soul's passions flowed out of the heart, exciting the body into action. To keep these lower souls from polluting the immortal soul in the head, the gods created another barrier in the form of the neck.

In *Timaeus* Plato built a spiritual anatomy with the brain at its apex. It would influence Western thought through the Renaissance, yet it was not powerful enough to bring the heart-centered school of Plato's own age to an end. In fact, Plato's most celebrated student, Aristotle, rejected the head and put the heart at the core of his philosophy.

For Aristotle, the brain did not square with his conception of the soul. In his philosophy, every object has a form, which can change as the matter that makes it up changes. A house emerges when stones are combined into a certain form, and its form disappears when the stones are taken apart. There is no one pillar or capstone in which the house resides—its form is everywhere in the house and nowhere in particular. The soul, Aristotle reasoned, is the form of living things. It therefore encompasses everything that a living creature does to stay alive. And since different organisms have different ways of life, they must have different souls, Aristotle concluded, each with its own set of faculties or powers.

To classify souls, Aristotle became the world's first biologist. He dissected everything from sea urchins to elephants, and while he didn't break the taboo on human dissections, he probably dissected stillborn babies. Aristotle tracked endless details of natural history, noting which species were warm-blooded and which were cold, which cared for their young and which abandoned their eggs. He found that he could classify species according to the faculties of their souls, ranging them on a ladder from low to high. At the bottom Aristotle put plants, because they had only vegetative souls that did nothing more than let the plants grow, heal themselves, and reproduce. Animals were placed higher than plants, because their souls had sensitive faculties as well: animals can see, hear, taste, and feel; and they can swim, fly, or slither. Humans stood alone at the apex of the natural world, with a rational soul equipped with faculties including reason and will—what we would call a mind.

Like the form of a house, Aristotle's rational soul was both nowhere and everywhere in the human body. Yet Aristotle also believed that specific parts of the body carry out its faculties. He scoffed at the idea that the brain could be such a place, since he saw from his dissections that many animals had no visible brain at all but still could perceive the world and give rise to actions. The brain itself could not have looked very impressive to Aristotle. Without freezers or formaldehyde to halt its decay, a brain quickly takes on the look and feel of custard—hardly the stuff of reason and will.

The heart, on the other hand, seemed to him to be a far more logical place for the rational soul's faculties. It is at the center of the body, and it was the first organ that Aristotle could see taking shape in the embryo. Greeks believed that the heart supplies life-giving heat to the body, and Aristotle saw a connection between heat and intelligence. Just as animals had more or less soul, they had more or less heat, mammals being warmer than birds or fish, and humans—he believed—the warmest of all. Unaware of the nerves, Aristotle imagined that the eyes and the ears were connected not to the brain but to blood vessels, which carried perceptions to the heart. These connections allowed the heart to govern all sensations, movements, and emotions. The brain, he wrote, simply "tempers the heat and seething of the heart." The big brains of humans are not the source of their intelligence, Aristotle argued, but vice versa: our hearts produce the most heat, which means they need the biggest cooling system.

It was not until a few years after Aristotle's death, in 322 B.C., that Greek anatomists emerged who were skilled enough to challenge him. In the city of Alexandria, the physicians Herophilus and Erasistratus overcame the ancient taboos and dissected hundreds of human cadavers, describing dozens of body parts for the first time, from the iris to the epididymis. Their most important achievement was the discovery of the nervous system. Earlier physicians had assumed these slender pale cords were tendons or the tips of arteries, but Herophilus and Erasistratus recognized for the first time in history that these fibers formed a distinct network that sprouted from the skull and spine.

They tried to make sense of this new nervous system in accordance with the ideas of their age. They believed that each breath carried a bit of the world-soul into the body, where it behaved just like water in a pipe. It flowed into the heart and through the arteries, bringing life to the body, some of it traveling to the brain. Herophilus and Erasistratus also discovered chambers in the middle of the brain—the ventricles—which were the only logical place for spirits to flow. Herophilus declared that these empty spaces house the intellect. From the ventricles, he believed, the spirits flow into the hollow nerves and out to the muscles, which they make bulge and move. The brain itself, he thought, had no command over the body, and even the spirits had limited power: the body's organs could move thanks to their own natural desires.

It would take another four hundred years for someone to match the anatomical skill of Herophilus or Erasistratus. In A.D. 150, a young doctor named Galen traveled from Turkey to Alexandria to immerse himself in their teachings. He studied the human skeletons preserved at their schools and read their ancient works in the city's libraries. Galen himself couldn't dissect human cadavers, because Romans were even more appalled by the notion than the Greeks. And so when he returned to Turkey, he had to make do with glimpses of anatomy. As a doctor to gladiators, he could peer through the windows ripped open by tridents and spears. He dissected an animal every day. By the time he was thirty, Galen had created a new vision of the body by synthesizing Aristotle and Plato with the medicine of Hippocrates and his own observations. The result was so dazzling that when he moved to Rome, he became doctor to emperors.

Galen's medicine rested on the transformation of food and breath into flesh and spirit. In his system, each organ had a special faculty—a soul-like power—that carried out a series of purifications. The stomach had a faculty for attracting food from the mouth down the esophagus and another faculty for cooking the food, turning it into a substance called chyle, which passed into the intestines, the surrounding veins, and the liver. The liver turned this chyle into blood. In the process, Galen argued, the liver filled the blood with a nourishing force that later physicians came to call the natural spirits. From the liver, the blood was believed to flow to the heart, passing through its left side. Impurities were attracted into the lungs, and the purified blood traveled into the veins, to be consumed by muscles and organs.

Some of the blood that entered the heart had a higher calling, supposedly trickling through the heart's inner wall to the right side, where it mixed with air from the lungs and cooked in the heart's innate heat, turning red and becoming imbued with vital spirits. The pulsating arteries attracted this blood and delivered its life-giving powers throughout the body.

The vital spirits that flowed up to the head underwent a final round of purification. They entered a mesh of blood vessels at the base of the skull (which came to be known as the *rete mirabile,* or marvelous network), where they became animal spirits, capable of thought, sensation, and movement. From there they flowed into the ventricles. Galen claimed that the ventricles were spherical, roofed with vaults of flesh, and linked by canals, designed to be inflated by the swirling animal spirits. The brain pulsated, he thought, in order to drive the spirits out into the hollow nerves, where they were driven out into the body, carrying sensations and movement.

To treat his patients, Galen restored the balance of this flow of natural, vital, and animal spirits. For instance, an overheated stomach could drive a flood of phlegm out of the brain and into the rest of the body. If there was too much blood—the hot and moist humor—a fever would strike. Purging and bloodletting could bring the humors back to their proper places, as could special herbs. But Galen believed he had discovered much more than a way to heal people: he had established a philosophy of the soul. He declared he had found the physical underpinnings of Plato's trio of souls—the vegetative soul of the liver, responsible for pleasure and desires, the vital soul of the heart, which produced passions and courage, and the rational soul of the head.

Galen came to understand the brain far better than anyone else in the ancient world, but he was not a modern neuroscientist disguised in a toga. What we think of as the brain was to him nothing but a pump, while human intelligence was lodged in the empty spaces of the head. Moreover, that intelligence was not unique to humans but also shared by the sun, moon, and stars. In fact, their heavenly bodies were so much purer than our own that their intelligence must be far superior, able to reach down to Earth to influence human affairs. For Galen, the animal spirits swirling within us were only tiny eddies in an ocean of purpose, intelligence, and soul.

—

In the centuries after Galen's death, around 199, his medicine was absorbed into the doctrines of Christianity. The early church fathers turned to him because they needed some new ideas about the brain and the soul.

According to the Old Testament, the soul is simply life itself, residing in the blood and disappearing at death. Christianity, on the other hand, anchored itself to a different sort of soul, an immortal one that faced eternal salvation or damnation. In Galen, the church fathers found a solution to this contradiction. The Old Testament soul became Galen's lower souls of the liver and the heart. The immortal soul had no physical dimension, but the church fathers put its faculties in the empty ventricles of the head, where they could not be corrupted by weak, mortal flesh. They even went beyond Galen to assign the front ventricle to sensation, the middle to understanding, and the rear to memory. The brain itself was merely a pump, squeezing the spirits out of the ventricles and into the nerves.

Galen's anatomy was not the only Greek idea that influenced Christianity, however. Many philosophers in Rome didn't accept Galen's claims about the brain, still preferring Aristotle's theories about the heart. They liked to point out how speech came out of the chest, which meant that the heart must be its origin. To refute them, Galen gathered together physicians, philosophers, and politicians of Rome to watch him silence the roaring lions of the Coliseum by squeezing their vocal nerves. But he did not manage to silence his opponents. As a result, the Christian heart became not only the seat of the passions, but also the site of moral conscience, an organ with powers of perception beyond the senses. It is no coincidence that Jesus is often pictured with an open heart but never an open brain.

After the fall of Rome in 476, the church lost touch with its Greek origins. Not until the twelfth century did European scholars rediscover Greek philosophy through their contact with Arabs. It took a long time for Europe to become reacquainted with the likes of Aristotle and Galen. The few surviving fragments of their works had been translated into Arabic, and the Arabic translations were then translated into Latin, getting encrusted with misreadings along the way.

Many Christians were suspicious of Greek ideas that seemed to challenge the church's teachings on the soul. Most heretical of all was the notion that the world was a void inhabited by atoms, invisibly small, indestructible particles of different shapes and sizes—twisted, round, bent, rough, and hooked. Atomists would say that the brain is not in and of itself cold; blood is not in and of itself warm. Those qualities, along with all others, emerge from the interaction of atoms that make them up. Moving through the cosmos without supervision or purpose, atoms cluster in countless different ways, producing an infinity of worlds. Epicurus, the leading atomist philosopher of Hellenistic Greece, believed that the gods were indifferent to human affairs; the world carried on thanks only to the jostling and mixing of invisible particles.

Epicurus also believed that the soul was no different from the rest of the cosmos; it was made of atoms concentrated in the chest. As these atoms leaked from the body, he claimed, they were replenished with every breath. Death came when the atoms of the soul suddenly fled the body, taking life with it. "Death is therefore nothing to us," the Roman atomist Lucretius wrote, "and does not concern us at all, since it appears that the substance of the soul is perishable. When the separation of body and soul, whose union is the essence of our being, is consummated, it is clear that absolutely nothing will be able to reach us and awaken our sensibility, not even if earth mixes with sea, and sea with heaven."

Christian theologians were outraged at the idea that the world required no purpose or providence. Dante spoke for many when he consigned Epicurus to the sixth circle of hell. "In this part Epicurus with all his followers, who make the soul die with the body, have their burial place," he wrote.

Aristotle, on the other hand, enjoyed a warm embrace. The thirteenth-century theologian Thomas Aquinas found foreshadowings of Christianity in his philosophy. Rejecting the blind battering of atoms, Aristotle saw everything in the cosmos coming into existence for a purpose. Aquinas simply revised the purpose, making it God's plan as laid out in the Bible. Aristotle's cosmology agreed with Christianity as well. He placed the Earth at the center of the cosmos because the element earth moved there naturally. The Earth was a realm of change and decay, surrounded by celestial orbs moving in perfect circles. To Aquinas, the imperfection of this lower world was the result of the fall of man, and the perfection of the stars reflected the heaven where blessed souls went after death. Man was both fallen and at the center of creation.

Aquinas also believed Aristotle's conception of the soul was consistent with Christianity. He opted for Aristotle's idea that the soul is the form of life, whether that life is plant, animal, or human. The human soul was not simply the form of the human body but a spiritual substance as well—one that survived death. Aquinas did not follow Aristotle blindly, however: he placed the soul's faculties, such as memory and imagination, in the ventricles of the head. But he made it clear that no physical organ could produce self-awareness or any other human thought.

Aristotle allowed Aquinas and his fellow friars to forge a new intellectual tradition, which came to be known as natural philosophy. They used reason to demonstrate that there was only one God, that He alone had created the world, and that His supreme goodness was evident in the workings of the world. As universities began to spring up across Europe in the 1200s, natural philosophers took control of them and set the intellectual tone of the continent.

These natural philosophers also revived Galen's anatomy, building anatomical theaters where medical students, philosophers, noblemen, and assorted gawkers watched surgeons peel away the skin of executed criminals, while anatomists sat in elevated chairs, reading aloud from Galen's books. European anatomists did not try to learn anything new during these dissections. In the words of one fourteenth-century anatomist, they were simply supposed to allow "the might of God to be marveled." Anatomists pointed out how admirably the body was assembled, with the soul's faculties housed in the three centers of the body—the liver, the heart, and the head. They described how the visible anatomy they exposed was traveled by invisible spirits, which the church taught were the immortal soul's tools for producing life.

Although they cherished an ancient philosophy, these anatomists wound up doing something altogether new. They were willing to dissect human corpses and could therefore enjoy a privilege that Galen had never had. They could see things about which he only speculated. It would take centuries of these dissections before anyone realized that Galen's teaching had not been based on experience with human tissues and organs. In the meantime, Galen's word became gospel in both anatomy and medicine. European doctors pored over translations of his work to learn how to balance the humors with herbs and bloodletting. Seeing themselves as natural philosophers, they trained for years in logic, grammar, and Greek, leaving the bloody work to surgeons.

In 1537, a twenty-three-year-old anatomist named Andreas Vesalius recognized that Galen was not perfect. In charge of the University of Padua's lectures on surgery and anatomy, Vesalius stepped down from his lecturer's chair to show his students the finest details of anatomy in human cadavers. Once, as he described how to bleed a patient, he made a quick sketch of the veins, after which his students begged him to make sketches of the arteries and nerves as well. Vesalius created a series of giant charts far better than any of the crude diagrams that had been available before. His fame spread as the pirated copies multiplied across Europe. The judge of Padua's criminal court started sending him a steady stream of cadavers from the gallows.

Vesalius began to suspect that there was something systematically wrong with Galen's work. It dawned on him that for all Galen's references to human anatomy, the old Greek doctor had never actually dissected a human corpse. Galen's womb belonged to a dog, his kidneys to a pig, his brain to a cow or a goat. All told, Vesalius found two hundred pieces of animal anatomy in Galen's human being.

Vesalius revealed his discovery to throngs of medical students, rigging up skeletons of humans and of Barbary macaques side by side. Peeling away the skin of human cadavers, he exposed more details that didn't fit Galen's claims. When he pointed out that some of the veins Galen claimed to be in the rib cage were not there, the professors in his audience interrupted him. They protested that Galen said the veins did exist. "Show them to me," Vesalius replied. The professors did not answer, seeing no reason why they should. Galen's authority was superior to Vesalius's eyes and their own.

Vesalius decided that he had to redo Galen's work completely, basing it on humans rather than animals. He worked with the finest block cutters in Venice and with draftsmen from Titian's workshop. The blocks were taken over the Alps to Switzerland, where they were assembled into a book entitled *De humani corporis fabrica libri septem* (*Seven Books on the Structure of the Human Body*). It was a magnificent atlas of his new anatomy, filled with men and women standing stripped of their skin and skeletons leaning lazily against columns in a rolling Italian countryside.

His book made Vesalius the most famous physician in Europe. But as revolutionary as it was, it had its flaws, many of which lay in his pictures of the brain. Vesalius dissected the brain not by removing it intact from the skull but by sawing slices off the top of the head, exposing fresh surfaces one after another. The farther down he went, the more mangled and decayed the brain became. And yet despite his blurry view of the brain, he managed some startling conclusions. He looked for the three spherical ventricles of the church's official anatomy and found a strange maze of horns and recesses. He searched for the miraculous net that supposedly transformed vital spirits of the blood into the animal spirits, and found none in humans. (Galen had seen them in an ox.) Vesalius wondered whether animal spirits actually flowed through the ventricles at all. Perhaps the flesh of the brain was a better place to look for powers of the soul. But he pushed this notion no further, partly out of fear.

> Lest I come into collision here with some scandalmonger or censor of heresy, I shall wholly abstain from consideration of the divisions of the soul and their locations, since today . . . you will find a great many censors of our very holy and true religion. If they hear someone murmur something about the opinions of Plato, Aristotle or his interpreters, or of Galen regarding the soul, even in anatomy, where these matters especially ought to be examined, they immediately judge him to be suspect in his faith and somewhat doubtful about the soul's immortality. They do not understand that this is a necessity for physicians if they desire to engage properly in their art.

—

In 1600 Western ideas about the soul were still guided by Galen, no matter what a few people like Vesalius thought in private. The souls of heart and liver still governed the emotions, desires, and appetites. The rational soul's faculties still swirled mysteriously in the void of the ventricles. The four humors governed not just physical health but the temperament as well. People born with an abundance of phlegm were indolent and dull. Blood made people bold, merry, oversexed, lucky, and gullible. Yellow bile made men hasty, envious, cruel, and unlucky. And black bile—also known as melancholy—produced not only sadness but thoughtfulness and detachment from ordinary life. If a patient built up too much of any humor it threatened not only to make him sick but to alter his personality. Too much black bile could turn harmless melancholy to delirium or violent insanity.

The four humors made the mind comprehensible even in madness. They linked each person to the four elements that made up everything on Earth, to the stars, and to the spirit world (even if that world included demons that could possess people and make them flail and rave). To cure madness, a healer had to restore patients to this divinely ordained balance of life.

One of the best surviving examples of this blend of medicine and religion comes from the casebooks of Richard Napier, who worked as both a minister and physician in Buckinghamshire in the early 1600s. Over forty years, Napier amassed sixty volumes of medical notes, documenting over sixty thousand cases ranging from plague to pimples. Among the patients who came to him were people who trembled, swooned, or thought they had rats gnawing their stomach or mice running inside their head. Napier (who himself was "much afflicted with mopish melancholy") diagnosed their madness by drawing their horoscope to determine how the stars and planets affected their humors. He believed the Devil possessed some of his patients, making them hallucinate or try to kill their family. (Sometimes the possessions were mild. One Edward Cleaver complained that when he finished thanking the Lord for his supper, "an ill motion came through his mind, saying 'Kiss my arse.'") In some cases Napier conjured the archangel Raphael to reveal whether his patients were bewitched.

With Galen's guidance, Napier bled his patients with leeches, fed them laxatives made of aloe and hellebore, and gave them tobacco to make them vomit. He used horoscopes to determine when his patients should take his drugs and gave them amulets stamped with planetary signs to wear around their neck. In the process of treating his patients, Napier drew out their stories, comforted them with religious sermons, and prayed with them for a cure. People came to his village from hundreds of miles away to be healed—a sign of just how widespread his conception of the mind was.

But the truce between Greek and Christian thought that shaped Napier's medicine was about to come to an end. Throughout the 1500s, scholars had been translating many ancient works from the original Greek for the first time. Old friends no longer seemed so trustworthy. Even Plato, who had given the church some of its earliest doctrines, wound up on its list of prohibited books. A new generation of philosophers inspired by Plato declared that the human soul could be influenced by the soul of the world, itself a living thing. Just as the human soul had spirits to carry out its will, the world-soul used spirits of its own, which it channeled through the planet to steer events on Earth. The planets influenced the human soul through a cosmic sympathy, just as a plucked lute string could make another string vibrate. Plato's followers began trying to harness the power of the stars with songs and other rituals, a practice they called natural magic. To conservative theologians, all of this smacked of pagan nature-worshiping that denied the power of God.

Even Aristotle could now inspire dangerous ideas. At the University of Padua in the early 1500s, the philosopher Pietro Pomponazzi claimed that Aristotle did not actually believe that individuals had immortal souls. He argued that if the soul was the form of the body, it could exist only within the body and must die with it as well. He found this consoling, not terrifying, and wrote that a man who knows his soul is mortal "will always be prepared to die. Nor will he fear death, since fear of the inevitable is vain; and he will see nothing evil in death."

Rome condemned Pomponazzi and warned philosophers against questioning the immortality of the soul. The Vatican made it official dogma, demanding that philosophers demonstrate the soul's immortality through natural reason. The result was a flood of attacks on Pomponazzi and the beginning of an intellectual struggle that would last for over a century. Along the way, philosophers would discover even more troublesome questions about the soul. They would find some of the most troublesome ones in an unexpected place: in the heavens themselves.

Chapter Two

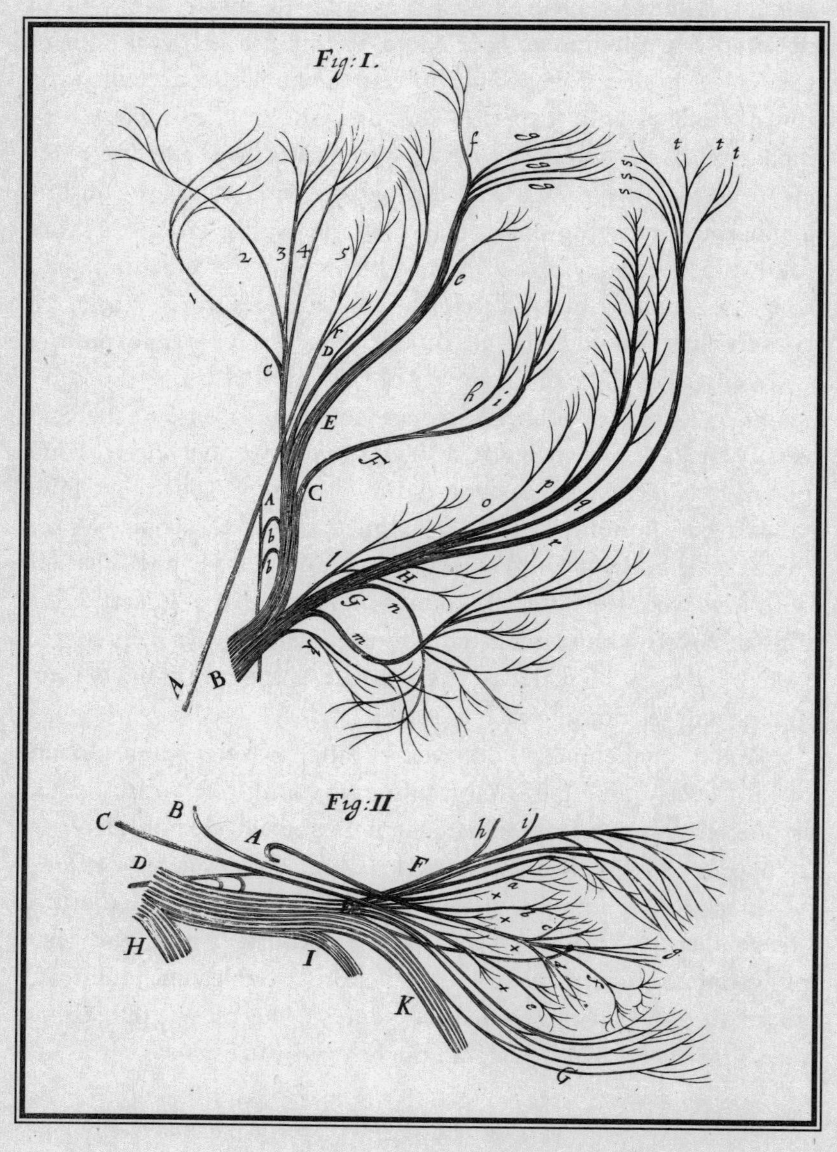

FACIAL NERVES, FROM
THE ANATOMY OF THE BRAIN AND NERVES.

World Without Soul

Fifteen forty-three was a watershed year for two kinds of anatomy. In the year in which Vesalius published his *Fabrica* and overturned the understanding of human anatomy, the Polish churchman Nicolaus Copernicus published a book that altered the anatomy of the world. The Earth was not the heart of the cosmos, Copernicus said, but just another planet revolving around the sun. Copernicus's impact would travel far beyond astronomy in the decades that followed. It would even help prepare the way for a new conception of the brain and the soul.

Both Vesalius and Copernicus were driven by dissatisfaction. Vesalius was dissatisfied with Galen, while Copernicus's dissatisfaction lay in Aristotle's picture of the cosmos. According to Aristotle, the Earth was at its center, surrounded by a set of rotating spheres that contained the planets and the sun, surrounded in turn by the orbs of the stars. Aristotle knew very well that planets do not move through the sky in perfect circles, and to account for their wandering paths, he installed fifty-five planetary spheres in the cosmos that together could produce their apparent movements. Yet Aristotle did not show that these spheres existed by making predictions about where the planets would go. Mathematics was the sort of minor detail he couldn't be bothered with.

The Greek astronomers who came after Aristotle were less interested in the physical nature of the cosmos than in making predictions for horoscopes and calendars, using whatever mathematical tricks they needed without pondering what those tricks might imply about the workings of the world. The most influential of these astronomers, Claudius Ptolemy, reworked the circular paths of the planets. He added epicycles to their paths that let him make better predictions; by nudging the Earth slightly away from the center of the cosmos, he could get better predictions still.

European scholars found themselves in a quandary when they discovered Ptolemy in the twelfth century. While his astronomy worked brilliantly, it didn't mesh with Aristotle's philosophy, which had become church doctrine. A debate dragged on for two centuries about how to reconcile the two. The church didn't persecute the astronomers for contradicting official dogma, however, because the astronomers never claimed that they were creating anything more than a mathematical description. They were not questioning the ultimate physical causes.

When Copernicus began studying astronomy in the early 1500s, it seemed to him that Ptolemy's model of the cosmos had grown into an irredeemably ugly hodge-podge. Like any good scholar of his age, Copernicus did not set out to think of a new system of his own but looked for an alternative explanation among the Greeks and Romans. He discovered that some ancient philosophers had argued that the Earth in fact moved. Given an excuse to rethink the cosmos, Copernicus concluded that the Earth was moved along with all the other planets in circles around the sun. He found that he could predict the positions of the planets just as well as Ptolemy and do it with a far more elegant model. The cosmos had been like a hideous monster, he claimed, but now its body was perfect.

For years after Copernicus's death the church took little notice of his new anatomy of the sun and planets. It was just another mathematical procedure that did not rise to the level of natural philosophy. Aristotle had explained why the stars and planets revolved around the Earth; Copernicus had no new physics to prove otherwise.

Toward the end of the sixteenth century, the Italian mathematician Galileo Galilei began to uncover that new physics. In Padua, the city where Vesalius had challenged Galen, he dared to challenge Aristotle with mere mathematics and measurements. If Earth really did strive to reach the center of the cosmos, as Aristotle claimed, the larger the object the faster it should fall, because it had a greater urge to reach its natural place. Galileo demonstrated that this was not true, that heavy and light weights both fall at the same rate. Galileo's new physics governed not only the Earth but the heavens as well. He pictured a ball rolling on a perfectly flat table on Earth, traveling forever; extend the table around the world, and the ball would move in a circle. Perhaps the planets had a circular inertia as well that made them revolve around the sun without end.

When Galileo wasn't performing experiments, he was searching for any observation he could use in his fight against Aristotle. In 1604, he was delighted to see a new star flare up in the constellation Sagittarius. The supposedly unchanging heavens had been dappled. Four years later, Galileo built himself a telescope and discovered that the moon was pocked with craters and encrusted with mountains, that shadows on the face of Venus moved just as Copernicus had predicted, that a private kingdom of moons orbited Jupiter, and that the Milky Way was made of millions of stars. Copernicus had found not a convenient mathematical short-cut, Galileo recognized, but the physical reality of the world—a reality in which the Earth was not the center of the cosmos, and in which the heavens were not a realm of perfection.

The new world Galileo was revealing was both confusing and frightening. In his 1621 poem "The Anatomy of the World," John Donne fretted that "new Philosophy calls all in doubt," declaring of the world, "'Tis all in peeces, all coherence gone." The Catholic Church did not want to see all coherence gone, putting Copernicus on its index of banned books in 1616. In 1624 a thousand people crammed into a lecture hall in Paris to hear three men offer forty propositions contradicting Aristotle, but before they could speak, the religious authorities forced the audience out of the hall and sent the speakers into exile. To teach a doctrine that defied the ancients could mean a death sentence.

A few monks and priests thought the church should take another path. It had to part ways with Aristotle. They saw no room in his philosophy for Christian miracles, a hierarchy of demons and angels, a first Adam, a genesis of the universe, or even for an immortal soul. God's providence was irrelevant in Aristotle's world made of self-transforming matter. If the church clung to him, it would be helpless to fight the new heresies seething in European cities in the early 1600s, from mystical nature-worshiping sects to skeptics who believed that nothing could be known for sure.

The headquarters of this assault on Aristotle was the cell of a Parisian monk named Marin Mersenne. Mersenne decided that the only way for the church to survive was to take the soul out of nature. God could rule only over a universe made up of soulless, passive matter that obeyed the laws that He devised or that was directly controlled by Him. The world, in other words, had to become a machine. Mersenne admired Galileo for using mathematics to discover the world's mechanical harmony, and he used the same methods himself to drive the magic out of music. A plucked lute string did not make another string vibrate thanks to a sympathy between their souls but thanks only to the vibrations of the intervening air. Mersenne kept abreast of developments in astronomy, medicine, mathematics, and every other branch of science, writing thousands of letters to correspondents as far away as Syria, Tunisia, and Constantinople. At home, he arranged debates in his cell, where some of the best minds in Europe revived all sorts of old heresies and reconciled them with Christianity.

One of his associates, a priest named Pierre Gassendi, embraced atoms, arguing that everyday experience confirmed that they existed—even in paving stones, as they are worn invisibly away by hooves and feet. He saw the universe as composed of indivisible, indestructible, invisible particles wandering through a void. Each kind of atom had an intrinsic size, shape, and weight that allowed them to form associations Gassendi called molecules, which took on new properties of their own. Gassendi claimed that salt could dissolve into water because its molecules were cube-shaped, as were the spaces between water molecules.

Gassendi believed that the idea of atoms, far from being heresy, was in perfect harmony with Christianity. God created atoms at the dawn of time, giving them the qualities that launched them on a course that would play out His providential design, "all this to the degree that he foresaw what would be necessary for every purpose he had destined them for," he said. He even believed that Christianity could allow for a soul made of atoms. According to Gassendi, humans and animals alike have a soul that burns within the body and that is carried in the semen from one generation to the next, igniting a new soul constituted out of the atoms of the embryo. This sensitive soul, as Gassendi named it, encompassed the brain and the nerves. The nerves received sensations and impressed them on the brain, like writing on a blank slate. Although it was made of mere matter, it could think. In our ability to perceive objects, to reason and make judgments, we are no different from animals.

But Gassendi also firmly believed that no arrangement of atoms could reflect on itself or perceive something beyond the images supplied by the senses. And since human beings can clearly see beyond the concrete and into the abstract, so they must have another soul. This rational soul was immaterial, but not completely free of matter, at least in this life. According to Gassendi, it was lodged within the brain, where it depended on the sensitive soul to bring it images of the outside world.

Gassendi believed he had found a natural philosophy that did proper justice to the Christian soul. Since the rational soul was not made of atoms, it could not be subject to the decay and death of all things made of atoms. It was immortal, liberated from the physical laws that enslaved atoms, and shielded inside the brain from the swarming sea of particles outside the body. Buffered by the sensitive soul, it did not instantly respond to pleasure or pain as an animal's might but could move instead in a straight line toward God. Just as He had put the atoms of the universe in motion to carry out His will, God had designed rational souls to move toward Him.

—

In 1625 Mersenne welcomed a young dandy to his circle. René Descartes liked to walk through Paris dressed in taffeta, with a plumed hat on his head and a sword at his side. He gambled, danced, and fenced. But Mersenne could see that he was also a gravely serious, fiercely ambitious philosopher. Descartes was on a divine mission to transform all natural philosophy into a completely new science.

As a boy he had not seemed cut out for such stuff. He was small and sickly, with a dry cough that lingered until he was twenty. "All the doctors who saw me up to that time condemned me to die young," he later wrote. He was so delicate that when he was sent off to Jesuit school at age ten, his teachers let him lie in bed till late in the morning. He would spend that solitary time deep in thought, contemplating the philosophy the Jesuits were teaching him. They instructed him in Aristotle and Aquinas, of course, but they also wanted to prepare their students for the seventeenth century in all its uncertainty. Nearly a century had passed since Martin Luther had nailed his theses on the door of the cathedral church in Wittenberg, marking the birth of Protestantism. By 1600 much of northern Europe had converted. In France the Catholic Church responded with an Inquisition to root out heretics and a Counter-Reformation of its own. Descartes's teachers introduced him to the Catholic faith not simply as a series of rituals, but as a drama to be played out in his internal life.

Descartes's internal life was dominated by a deep-running Catholic faith and a nagging doubt about everything else. Only mathematics gave him the deductive sense of certainty he craved. For a time he studied law, but had no real appetite for it. He wanted to discover knowledge in experience itself rather than in books, and so, at age twenty-two, he became a soldier.

Traveling to the Dutch city of Breda, he joined the army of Prince Maurice in his battles against Spain. He arrived during a long lull in the fighting, but he was not bored. The army he had joined was a scientific one, full of telescope-makers and fortress-building geometers, and soon he became fast friends with a soldier-scientist named Isaac Beeckman. Under Beeckman's guidance, Descartes began a spectacular career as a mathematician. He found new, elegant solutions to problems that had vexed mathematicians since the ancient Greeks had first posed them. He figured out how to translate curves and lines into equations and then back again. With Beeckman's help, he also studied the strange new natural philosophy Galileo was building. Descartes admired how Galileo was using mathematics to understand the world, but he wanted to bring order to the entire undertaking. He set out to discover the fundamental rules for the search for knowledge.

After a year without seeing a battle, Descartes left Breda and Beeckman to find another war to fight. He traveled to Bavaria to serve under Duke Maximilian in his fight against Protestants, but when he arrived that war had turned to negotiations as well. To pass the time, Descartes settled into a rented room in the city of Ulm. He hardly left it for weeks as he contemplated the nature of knowledge, until his nerves finally began to fray.

One night he lay in bed, his eyes shut but fluttering. He felt a crippling fear, and the right side of his body became weak. He turned to his left side and dragged himself forward. He tried to straighten his body but found himself spinning on his left foot instead. Before him stood a church. He walked through its open gate into a courtyard, in the hopes of finding a refuge and a place to pray. But a wind rose and threw him against the church wall. A stranger appeared before him. He told Descartes to search for a Monsieur N. Monsieur N. would give him a melon from another country.

Descartes opened his eyes. For the next two hours he lay in bed, praying, at a loss for the meaning of the dream he had just had, until he fell asleep again. He heard a terrifying noise and he opened his eyes a second time. Sparks danced through the room. As he blinked a few times, Descartes drifted back to sleep and fell into a final dream. On the table of his room was a book filled with verse. A man appeared and asked him to find a poem; before Descartes could find it, both the man and the book had vanished.

In later years Descartes would look back on that night as the turning point of his life, a moment when he was given a vision of the purpose of his existence. For days afterward he tried to work out the meaning of the dreams. They were signs of God, he was sure. The melon must represent the solitude he had enjoyed since he had left France. The wind and the pain in his side were connected to an evil genius who was trying to force him into a place he wanted to go on his own. The terror of the second dream was his remorse over the sins he might have committed, and the thunder was a sign of the spirit of truth descending to him. The final dream showed his future: to search the book of life for certainty.

Out of this crisis Descartes emerged with his "doctrine of clear and distinct ideas." The way to uncover the truth, he decided, began with accepting nothing as true. He would tackle every scientific problem by dividing it into parts, solving the simplest parts first. In working through a problem, he would leave nothing out. Mathematics lent itself most easily to his method, but Descartes was convinced that his method would ultimately let him understand all of nature.

When Descartes returned to Paris in 1625, he slipped comfortably into Mersenne's circle. He shared their dream of interpreting nature through the motion of particles, replacing nature's souls with machines, and finding a certainty on which the Catholic Church could rest its faith.

He began his search for certainty not in the heavens or the Bible but in the nervous system. He followed perceptions from the external world into the mind, asking how we can be sure that what we are sensing is real. This was a new question. Before 1600, it would not have made sense. Aristotle believed that when he looked at a green mountain, its form traveled through the air and imposed itself on his eye. If conditions were good, Aristotle's eye saw the mountain accurately, because it actually contained the mountain's form. The form was then expressed in his soul, allowing him to experience the mountain's shape and its greenness. Galen gave Aristotle's argument a medical authority, claiming that the form struck the eye's jelly, which he thought was the soft tip of the optic nerve.

No one seriously challenged this account for 1500 years, until the close of the sixteenth century. The astronomer Johannes Kepler decided to study the eye in the same way as he would study a glass lens—as an optical instrument subject to the laws of physics. He scraped away the back of an ox eye and found that miniature images were projected on its inner wall. Those images were not the forms of objects but inverted pictures created as refracted rays of light traveled through the eye. Kepler found that each point of the object was perfectly transformed into a corresponding point in the image. From the retina, he argued, this inverted image traveled to the brain. How it was flipped right side up in the brain, however, he dared not speculate.

Descartes followed in Kepler's footsteps, but made an even more sophisticated break from Aristotle. Whereas Kepler had modeled the lens as a simple sphere, Descartes figured out how to study light as it was focused through the lens's true shape. He proposed that the color of light was a product of the speed at which a ray struck the eye. Humans got an accurate perception of a green mountain—and the rest of the world—thanks not to Aristotle's forms, but to the geometry of the human eye and the physics of light. Descartes came to believe that the very construction of our senses and nerves guaranteed his doctrine of clear and distinct ideas.

As he struggled to create his New Science, Descartes began to crave solitude, and in 1628 he returned to Holland. In his self-imposed exile, Descartes invented the *x-y* coordinates of the Cartesian graph, which turned space into a matrix of numbers. He dared to decode the rainbow, the sign of God's covenant, as nothing more than the bouncing of light within raindrops. Descartes could even predict the correct position at which rainbows would appear in the sky. Their color was not part of their form, but simply the result of an interplay between light, matter, and the eye. Descartes lifted his physics from the air to the universe itself, driving the planets and the stars through swarming whirlpools of invisible particles.

If he could account for the universe with matter in motion, there was no reason he couldn't do the same for the body. Descartes became a familiar sight at the butcher shops in Amsterdam, where he would buy freshly slaughtered animals. When visitors asked to see his library, he would take them into a room where he kept carcasses in various stages of dissection. "These are my books," he would say.

For all his boasting, though, Descartes didn't dispute Galen much at all. He even accepted the marvelous network at the base of the brain, despite the fact that Vesalius had shown that it didn't exist in humans. Descartes's brilliance lay elsewhere: he came up with an explanation of the workings of the body in the language he might use to describe a clock. He treated the body as an "earthen machine." A body could live and move, he argued, without the help of life-giving souls. "We shall have no more occasion to think that our soul excites the movements—those which we do not experience to be presided over by our will—than we have to judge that there is a soul in a clock which causes it to show the hours," he declared.

There was no need, for example, for a soul to reside in the liver in order to attract food from the stomach. It would be enough for particles of food to be broken up in the stomach and then jostled into the pores of the liver. From the liver, particles could travel to the heart, where Descartes believed they were transformed by heat into blood. This was not an Aristotelian heat that was somehow innate to the heart; it was just a fire without light.

Descartes agreed with Galen that the blood that rose from the heart to the head was transformed into animal spirits and entered the empty ventricles. But for Descartes the spirits were as material as blood and bones. He considered them "a subtle wind" made up of the strongest particles of the blood, which pushed their way into the ventricles. These spirits pressed against the walls of the ventricles like a stiff breeze against a sail and then surged into the nerves, which Descartes believed were hollow tubes. As the nerves expanded with spirits, they made muscles swell and contract. A swollen nerve could tug on the muscles at the sides of the eyes, turning our gaze left or right.

Descartes made the brilliant leap of recognizing that this tugging might take place without a tugger. Perception could turn to action without any soul intervening. Think of what happens when your foot gets too close to a fire, he said. The flames release particles that crash into the foot, tugging at the nerves in the skin. He speculated that there are fibers embedded in the walls of the hollow nerve tubes that run all the way up to the head, where they are attached to the walls of the ventricles. The fire yanks one end of each fiber like a bell rope so that its other end pulls open a pore in the ventricle's walls. Animal spirits in the ventricle rush into the opened pore and down the nerve to the foot. There they inflate a muscle, automatically causing the foot to yank itself away from the fire. The person doing the yanking has as little need for a soul as a mechanical doll.

Coughing, blinking, yawning, walking, breathing—Descartes recast all the movements of the body as tugs of nerves. When a corpuscle struck the eye, it simply sent rays of light to the retina, setting nerve endings in motion, and the disturbance traveled on to the brain. The images from each eye were combined into a single picture by the simple act of adjusting the eyes so that they both focused on the same object. Memories were caused just as mechanically. Once a nerve had finished opening a ventricle pore, it usually closed again. But if a pore were opened again and again by the same sensation, it might become wider than the other pores around it.

Descartes was convinced that animals were nothing more than intricately crafted machines made up of passive particles. Humans were different only in that they contained another substance utterly divorced from matter: the rational soul. Unfettered by mechanical laws, the human soul was capable of things no machine could achieve: consciousness, will, abstraction, doubt, and understanding. "There is only one soul in us," Descartes declared, "and that soul does not have in itself any diversity of parts." But how could a thing with no size, shape, or place control a material body? Descartes believed that the intersection had to be somewhere in the head, since that was where the animal spirits turned sensation into action.

To find the soul's hiding place, he carefully studied the brains of calves (despite the fact that calves supposedly had no souls at all). Descartes made woefully crude drawings of the brains and imposed on them what he had read in anatomy textbooks. He was drawn to a kernel of flesh the size of a rice grain called the pineal gland. Galen had claimed that the pineal gland was located at the point where the vessels of the marvelous network joined back together. It also projected into the empty space of the ventricles. Its location was perfect for Descartes, who argued that the animal spirits produced in the marvelous network were focused into the tiny space within the gland, where the rational spirits could perceive the outside world and then steer the animal spirits to do the body's work—to speak, to write, to walk.

Of course, the truth is that humans actually have no marvelous network at the base of their brain. The pineal gland is actually far away from where Galen claimed it was, but Descartes was hardly the first person to make these anatomical blunders. He was not interested in making an accurate anatomy so much as showing how easily Galen's body could be transformed into an earthen machine.

The pineal gland, Descartes asserted, was lashed by the most delicate fibers to the brain. That arrangement allowed the rational soul to make the gland quiver and shake—an operation so slight that even an immaterial thing could manage it. The soul's twitchings aimed the spirits at particular pores in the walls of the ventricles, which then traveled through the nerves and carried out the soul's will. The rational soul never stopped thinking, either in sleep or after death. Thinking, after all, was its essence, the definition of its existence. Descartes sometimes even claimed that the rational soul did not actually need a brain to think. "The mind," he declared, "can operate independently of the brain."

In 1632 Descartes began to combine all of his work—on the body and soul, on rainbows and clouds, on corpuscles and planets—into one grand book he simply called *The World*. But the following year his entire philosophical project was turned upside down when he learned that Galileo had been condemned by the church. For years Galileo had been trying to persuade Pope Urban VIII that it was not heresy to say that Copernicus was right, but Galileo's enemies eventually turned Urban against him. He was summoned to Rome to face the Inquisition, and ultimately he was forced to deny Copernicus. His books were banned, and he returned home to a life of house arrest.

The news made Descartes stop writing *The World* altogether.

"I wouldn't want to publish a discourse which had a single word that the church disapproved of," he wrote to Mersenne, "so I prefer to suppress it rather than publish it in a mutilated form."

Descartes needed a new strategy. He reworked the argument of *The World* from scratch. He decided he would have to anchor his work to a far deeper kind of certainty—a metaphysical one. To find it, he had to swim into an ocean of skepticism into which no one had even waded before. He would no longer claim that we could trust our senses because they were by their nature trustworthy. That might open him up to accusations that he depended on nature rather than God for his certainty. Instead, Descartes allowed that maybe everything he saw around him was an illusion. How could he know that anything at all existed? He retreated from the pageant of his senses to the one thing that had to exist by the very nature of the question: his own self. It was an undeniably clear, distinct idea. The entire universe now depended on the soul for its survival. Descartes summed up his newfound certainty with his famous motto *Cogito, ergo sum:* I think, therefore I am.

From those three words flowed a security that would bring him back to the natural world, and even the most suspicious priest would be carried along by the waves. Any idea as clear and distinct as the *Cogito* must also be true. Descartes's mind existed beyond a doubt, and it could imagine perfection. Perfection could not be imagined without its existing, he argued, because a cause could not be any less than its effect. Therefore God exists. And by God's very nature, He would not create an elaborate deception for Descartes. "He has not permitted any falsity to exist in my opinion which he has not likewise given me the faculty of correcting," he wrote. God had created the universe and the laws that governed it, and he had stamped those laws on our own minds. Here was a divine guarantee that if a person followed Descartes's methods of clear ideas, he could learn how nature worked.

Descartes turned this new argument into a short book he called *Discourse on Method*. Philosophical works were traditionally written in Latin, but Descartes wrote his in French, so that anyone, man or woman, might understand what he had to say. Descartes recounted his life as a lonely spiritual journey, making none of the customary references to Aristotle or other ancient philosophers, instead presenting his *Cogito* and laying out a few of its implications. He explained how animals could move, react, and survive without a soul or self-awareness and how humans were much the same but also had a rational soul, which directed the flow of animal spirits through the ventricles.

Descartes's sketchy introduction to his philosophy cleared the French censors in April 1636. The book was a sensation and word spread to other countries. Admirers tracked him down. Noblewomen took day trips to visit him, their adoration helping to turn his ambition to arrogance. Pretending that he had come to his methods completely on his own, he denounced his mathematical mentor, Beeckman, for claiming even the slightest influence on him. Descartes declared that his critics were nothing but flies, their letters good only for using as toilet paper. Over the next seven years, when he wasn't sneering at his critics, he was expanding the *Discourse* with two longer books, *Meditations* and *Principles.* He was now convinced that universities would soon be teaching his books rather than Aristotle's.

For all his care to avoid condemnation, Descartes wound up persecuted by Catholics and Protestants alike, much of his trouble coming from his disciples. His friends at Dutch universities began to teach his philosophy but without the metaphysical props that supported it. Riots broke out in lecture halls. Dutch theologians saw his philosophy as casting doubt on the Trinity, on the soul itself. They had him hauled into court as a Jesuit spy, threatening to expel him from Holland and to burn his books. In 1645, after a particularly bad year of accusations and scandals, Descartes wrote, "I have aged twenty years. . . . I feel weak and more than ever I need comfort and rest."

Things got so bad that Descartes decided to try his luck in France. But French theologians wondered how the miracle of transubstantiation could occur in a Cartesian world. To show that he was not an atheist, Descartes recast the miracle in terms of his new physics. He claimed that Jesus' soul joined with the bread just as the human soul joined with the body in the pineal gland. But his new account couldn't stop the accusations.

With no place to call home, in 1648 Descartes got an extraordinary invitation. Christina, the queen of Sweden, asked him join her court and teach her philosophy. The twenty-two-year-old queen was a notorious freethinker who cursed regularly, read Virgil during mass, and combed her hair once a week. At first Descartes was reluctant to go, but the queen gradually charmed him. By the time Christina's warship arrived for him, Descartes was giddy. He put his hair in ringlets for the voyage and dressed in long pointed shoes and white fur-trimmed gloves.

Once he arrived in Sweden, Descartes quickly realized he had made a mistake. Christina commanded that he arrive at her palace for her lessons at five in the morning, an hour not suited to a man who spent his mornings for over forty years meditating in bed. The kingdom was suffering its coldest winter in memory, and Descartes complained that in Sweden men's thoughts froze like water. He caught a chill on the first day of February 1650, and he went to bed with pneumonia. He dismissed his doctors, prescribing for himself wine flavored with tobacco to make him vomit out the cold phlegm that he concluded was afflicting him. For days he lay in bed, raving to God about the misery of man. But he came around and recognized that he was dying. Descartes spoke to his own soul as he felt his life fading. "My soul, you have been a captive for a long time," he whispered. "Now the hour has come when you must leave your prison, this body; you must bear this separation with joy and courage."

No one's soul would be the same after Descartes. He made the first great assault on the lower souls of the body, turning them into the clockwork of mechanical matter. The machinery of life might even produce sensation, memory, and movement. And while the notion of a rational soul, lodged in the pineal gland tugging at nerve fibers, existed only in Descartes's own imagination, he had set the study of the brain on a completely new course. Building on his clear and distinct ideas, the next generation of natural philosophers would launch the Neurocentric Age.

Chapter Three

THOMAS WILLIS.

Make Motion Cease

As René Descartes was cutting open calf brains in the Netherlands and trying to fit the soul into the pineal gland, Thomas Willis was still a young boy, walking to school each morning across the meadows of southern England. There were a few clues on those walks that he would grow up to become the world's first neurologist. He kept a sharp eye on the animals he passed and was struck by their intelligence. A newborn lamb knew immediately how to nurse. Chickens knew how to peck at grains of corn almost as soon as they emerged from their egg. Animals could also acquire more knowledge than they had been born with. Horses searched for lush meadows and remembered afterward how to find them again. A fox could get a turkey out of a tree by walking around and around the tree until the turkey got so dizzy that it fell to the ground. Perhaps Willis wondered whether the animals could think. And if they could, did they have souls?

These thoughts could just as easily have been signs that this boy walking through the meadows was destined for a quiet life as a clergyman. In the early 1630s, that certainly would have been the natural course for a boy as bright and pious as Willis. On his walks, Willis would give away his food to beggars he met along the way. "This boy will starve himself," his father said, and ordered him to eat his breakfast before leaving home.

Willis spent a serene childhood in a cradle suspended between the posts of church and king. A century had passed since Henry VIII had broken with Rome and made himself the head of the Church of England. England had become a Protestant nation in the decades that followed, and although the transformation had been tense and sometimes bloody, the country had avoided the full-scale religious wars that crippled the Continent. During Henry's reign, the actual experience of going to church hardly changed at all; it was his son, Edward VI, who first permitted clergy to marry, imposed the use of the Book of Common Prayer, and had images and altars removed from churches. But Edward was succeeded by his half-sister Mary, a devout Catholic, who reversed some of Edward's reforms and had hundreds of Protestants burned at the stake. When Mary died in 1558, another half-sister, Elizabeth, became queen. She repealed Mary's Catholic laws, reestablishing the Church of England, but did not swing all the way back to Edward's austere Protestantism. Instead, she managed to strike a careful balance that held till the end of her long reign. She crushed Catholic conspiracies to assassinate her, yet she also placated Catholic noblemen by retaining some vestiges of the old Roman rituals. She also compromised with Protestant reformers who were determined to purify the church altogether—people who came to be known as Puritans. While Elizabeth didn't meet their demands, she did let Puritans participate in Parliament and even set up their own congregations.

When Elizabeth died unmarried and childless in 1603, the throne passed to her distant cousin James Stuart. James was already king of Scotland and had been since he was an infant.

When he took the English crown, at age thirty-six, he declared himself an old, experienced king who could rule England, Scotland, and Ireland as a peaceful, united nation. For his coronation he had ninety-foot-high arches of wood and plaster built along his procession route, bearing pictures of England as an orderly, blessed nation—a "bower of plenty"—and passages from Virgil that promised the start of a golden age. James revived the medieval notion that monarchs are chosen by God and accountable to Him alone. Throughout his reign, James reminded his subjects of the divine right of kings. He liked to write hectoring pamphlets in which he declared, "Kings sit in the throne of God and thence all judgment is derived."

Puritans hoped that James, who had been raised in staunchly Protestant Scotland, would liberate the church, but he showed no interest in turning his new kingdom into a religious experiment. Instead he tried to herd all his subjects into one broad church organized by rank and order. To Puritans who wanted all the old hierarchy of Rome eliminated from England, he simply replied, "No bishop, no king." Dissidents were persecuted, and some Puritans began to flee to the English colonies in America.

Meanwhile, the Church of England grew more powerful, making Puritans even more bitter as they watched it enrich itself with a tenth of every year's harvest. Parliament also became exasperated by James's extravagant tastes. Queen Elizabeth's long war with Spain had left the royal coffers nearly empty and yet James put on grand masques and banquets. But even James's most hostile subjects did not question his right to rule. They simply wanted to rein him in.

In 1625, when Thomas Willis was four years old, James died and his son Charles inherited the throne. Small, stoic, and shy, Charles came to believe that he had become "God's lieutenant on Earth." But divine right did not bring Charles the political skills of his father. When he called Parliament for the first time, he was immediately attacked by the angry members. Just before his death his father had launched a miserable failure of a war against Spain.

Parliament demanded an accounting, and Charles haughtily refused. Puritans called on him to loosen the constraints of the Church of England. He responded by insisting that all churches use elaborate rituals that smacked of Rome. His favorite bishop, William Laud, enforced his will, telling the country that in order to obey God, they had to obey their king. Parliament responded by refusing Charles money. He demanded forced loans, which his own judges declared illegal. Charles responded by shutting down Parliament in 1629. It would not meet again for eleven years.

During that time Charles burnished his image as the father of the nation. He encouraged the old tradition that the king's touch could magically cure scrofula, a disease (known then as the King's Evil) that caused hideous swollen boils. Thousands of sufferers came to Charles's court and knelt before him so that he could lay his healing hands on them. Each time Charles did so, he demonstrated his divine right to rule. (And to avoid any confusion, it was made a crime for anyone other than the king to claim that he could cure the disease.)

Charles viewed himself as the rational soul governing a tranquil, united Britain. Court poets celebrated him and his wife, Henrietta Maria, as the spiritual axis of an orderly cosmos. The poet Thomas Carew called them

This royal pair, for whom Fate will
Make motion cease, and Time stand still,
Since Good is here so perfect as no worth
Is left for after ages to bring forth.

This comforting image dominated Thomas Willis's childhood. Although many parts of England were becoming staunchly Puritan, Willis's family belonged to a community of dedicated royalists scattered around Oxford. They would give their lives in defense of their king and the Church of England. And so it came as no surprise that at age sixteen Willis became a student at Oxford University.

For centuries, Oxford University had supplied the church with priests and monks. In the 1500s England's nobility began sending its sons in great numbers to study for a few years before they joined the court. When Henry VIII made England Protestant, Oxford became a nursery for the new Church of England. Puritan theologians also taught at the university throughout James's indulgent rule, but in the 1620s the university became more hostile to its Puritans. In 1629 they faced extinction when Bishop William Laud became Chancellor of Oxford.

Laud, who also became archbishop of Canterbury in 1633, made Oxford a microcosm of what he hoped to make the nation as a whole. Using the university's power over the town, he shuttered all but three of the three hundred alehouses and searched out the whores and thieves. At the university, he demanded that students tip their hats to their masters; those who failed to do so were fined, then flogged. Laud scrutinized every detail of college life, down to the exact length of haircuts. According to one college official, Laud's inspection "threatened to rival the siege of Troy."

Laud tolerated no dissent from his own lush, ornate vision of Christianity. The Puritans might think of a church altar as nothing but a wooden table for parish meetings, but to Laud it was "the greatest place of God's residence upon earth." He built up England's cathedrals in lavish—Puritans would say papist—luxury. To Laud, Rome might be a misguided church but it was a true one, while Puritans represented the Antichrist. They had turned God into "the most fierce and unreasonable tyrant in the world." At Oxford, he demanded that all masses be in Latin; vespers had to take place in all colleges simultaneously. Divine right and every other point of Laud's theology had to be taught to the letter. Anyone who preached on a forbidden subject was expelled from the colleges. The few Puritans willing to keep quiet survived at Oxford under his rule, but most were expelled or left in disgust. By the end of the 1630s, Laud had turned Oxford into a factory for producing loyal priests ready to preach the authority of church and king.

Laud did believe in learning, if it kept within tight constraints of tradition. Under his rule, Oxford astronomers began to teach Copernicus but did not stop teaching Ptolemy. They sailed to South America and Alexandria to study the stars with their own telescopes. Laud even appointed a professor of Arabic to mine Muslim literature for mathematical manuscripts. But he made sure that students' education was rooted in the medieval traditions of rhetoric, logic, and moral philosophy. Natural philosophy would be taught only from the books of Aristotle, whose authority was supreme. To earn their degrees, students at Oxford had to present disputations—ornately constructed Aristotelian arguments—on cutting-edge subjects such as whether summer or winter presents greater delights.

Willis got an education stamped with Laud's mark. But when he wasn't in the lecture halls, he glimpsed a scandalous form of knowledge on which his professors frowned. He got it thanks to his low birth.

Students who came to Oxford were graded by their social class. The highest were called fellows, followed by fellow commoners, scholars, commoners, and, last of all, servitors. Willis was a servitor, which required him to work for a member of the university to pay his way. (While fellows and scholars enjoyed finely cooked meals at their own tables, servitors had to fight over the scraps that were left behind.) Willis found work with Thomas Iles, a canon of Christ Church, serving Iles his meals, lighting his fires, fetching his wood, and making his bed. He also worked for Mrs. Iles, whom an Oxford diarist named Anthony Wood called "a knowing woman in physic and surgery" who "did many cures."

At the time, it was common for someone in Mrs. Iles's position to practice medicine. In all of England, there may have been a few hundred university-trained, licensed physicians, leaving the vast majority of English people to find unofficial healers for their care. If you lived in Oxford and needed a broken bone set, a locksmith on Cat Street could do it for you. At the markets, you could buy medicine made by self-taught empirics. Outside London, women played the biggest role in medicine. Midwives delivered babies, while wise women offered magical cures. Housewives had to deal with most of the medical disasters in their own home, and many kept notebooks of their favorite remedies. A kitchen was like a clinic, full of tooth-pulling forceps, lancets for bleeding, suture needles, and even alembics for distilling remedies. Some women took it upon themselves to treat their entire communities, establishing little drug factories in their home.

Thomas Willis made remedies for Mrs. Iles by drawing not on Galen but instead on popular medical recipe books of the time, such as the *London Pharmacopoeia*. Willis's professors at Oxford might teach him an orderly system based on Aristotelian logic, but in these encyclopedias he encountered an unruly underground culture of ideas that had been thriving and spreading throughout Europe for centuries. Here he found not the cool, logical systems of the Greeks but a mishmash of remedies that included the work of opponents of Galen—the alchemist-healers.

—

When Europeans began to look over the shoulders of Arab scholars, around 1100, one of the things they discovered was an ancient art for transforming matter. From Egypt to China, people had learned how to manipulate nature, whether to extract metals from ores or create dyes for clothes. Their art came to be known as alchemy. The early alchemists found that certain substances could transform other substances. They gave them names and divided them into categories. They tried to uncover the underlying reality that made it possible to transform one to another. Following Aristotle, they believed that metals could be transformed into other metals by changing their form. In Egypt and the Middle East, alchemists became convinced of the existence of a substance with a transforming power above all others, a substance that could turn all base metals into gold. The philosopher's stone, as it was called, became the obsession of alchemists for centuries.

Europeans learned about alchemy from the Arabs and began to

practice it as well. But unlike logic or anatomy, alchemy was never embraced by the medieval universities. Some alchemists worked as craftsmen, using alchemy for ordinary business such as making fake amber out of turpentine or pearls from seashells. But many were on a secret, spiritual quest. They wanted to find the philosopher's stone and unveil nature's deepest secrets. At the same time, they wanted to keep their knowledge from getting into the wrong hands, and so they wrote about their work in codes and allegories. Alchemists believed they could find the philosopher's stone only if their spirit was pure. Christian mystics even saw in alchemy an alternative to the Aristotle-influenced theology of the church, a promise that the world could be transmuted back to Eden.

Alchemy was doomed in its ultimate goals, but it nevertheless helped give rise to modern science. As obscure as its recipes could get—"cast a corpse into the salty sea and have a rooster devour a fox to prepare red dragon's blood"—you can still reproduce them today if you know what compounds the corpse and the rooster and the other symbols stand for. While they never found the philosopher's stone, the mystical quest of alchemists led them to many great discoveries along the way.

Alchemy promised to make men not just rich but healthy. The same elixirs that alchemists believed would make base metals noble they claimed could cure all kinds of ailments. They studied the healing properties of spa waters, discovering the minerals lurking within them. Still, most respectable physicians ignored medical alchemy and held fast to Galen. Galen's medicine was based on bleeding and purging, applying blistering plasters, and prescribing medicines from plants and animals—not on forcing strange metals and salts down people's throats. Medical alchemy languished in the shadows for centuries, until June 24, 1527. On that day in Basel, Switzerland, a crowd of rowdy students from the university gathered around a drunk obese doctor named Philippus Aureolus Theophrastus Bombastus von Hohenheim. As the students cheered, he put a torch to a pile of books by Galen and his Arab interpreters. Von Hohenheim had declared war.

Von Hohenheim was better known by the name Paracelsus—Latin for "Equal to Celsus," referring to one of ancient Rome's greatest physicians. His name reflected his arrogance. At age fourteen he set out to find a teacher worthy of his respect, but in Basel, Tübingen, Vienna, Wittenberg, Leipzig, Heidelberg, and Cologne he found only fools. "They are vainglorious babblers in all their wealth and pomp," he declared, "and there is not more in them than in a worm-eaten coffin." Paracelsus grew up in Swiss mining villages, and the miners taught him more than philosophers. They showed him how metals grew in the Earth, how men who went underground grew ill and unable to breathe. As Paracelsus wandered Europe, he gathered knowledge from anyone he thought was worth listening to—"old wives, gypsies, sorcerers, wandering tribes, old robbers," he wrote. He worked as an army surgeon in the Netherlands and Italy, was taken prisoner in Russia, and made his way from Ireland to Istanbul.

Paracelsus returned home at age thirty-three. He had become not simply a physician but an alchemist and a mystic, which to him were all one and the same. The entire universe was an alchemist's crucible, in which God had created all nature through a chemical separation. But for Paracelsus, Aristotle's four elements were not the fundamental building blocks of matter. He believed instead in three entities he called principles: salt, sulfur, and mercury. They conditioned the matter they inhabited. Salt was what gave every object its solid state. Sulfur was responsible for an object's burning. Mercury produced the smoke when it burned.

But along with dull matter, Paracelsus also saw the cosmos seething with spiritual forces—angels, witches, monsters, and beings he simply called "archei": souls that governed the stars, the metals of the Earth, and even parts of the body. Paracelsus believed that souls could be in sympathy with or antipathy to one another, attracting or repulsing across great distances. He declared that the only way to learn about nature was through a divine union with the soul of the thing being studied. He believed that natural magic—control of the hidden archei—could reveal the universal sympathy of like to like, of microcosm and macrocosm. By reading comets in the sky, Paracelsus predicted the overthrow of kings and the triumph of Christ. His prophecies circulated throughout central Europe as hugely popular pamphlets.

As a doctor, Paracelsus saw an archeus in every organ, acting as an internal alchemist, transmuting food into flesh and bone. Each organ was linked to a heavenly body—the sun to the heart, the seat of the soul and life, and the moon to the ventricles, where reason was located. Diseases likewise were caused by foreign souls invading the body. Mania was a disease like any other, one that Paracelsus believed could be cured by cutting open the fingertips to let the disease escape from the body. For other illnesses, he prescribed metals or plants whose souls corresponded to the soul of the disease. A physician had to learn the language of nature to master medicine. The eye-shaped flowers of eyebright plants were signs of their sympathy, for example, indicating that they cured bad eyesight. Paracelsus worked as furiously in his laboratory as any alchemist searching for the philosopher's stone, but instead of gold, Paracelsus searched for metals that could be used for new medicines. Alchemy, Paracelsus claimed, would bring us back to the perfect state of health we enjoyed in Eden.

As remote as Paracelsus's ideas sound today, he made some tremendous advances in medicine, such as inventing laudanum, a mixture of alcohol and opium that served as the standard painkiller in Europe for more than four hundred years. He was also the first physician to see diseases as particular things caused by particular external influences that affected particular parts of the body, and that could be classified and treated. Paracelsus might have seen spirits where we now see bacteria or asbestos fibers, but the fact remains that he was able to recognize and describe specific diseases with far more precision than other physicians of his time. In Paracelsus's day, a miner's trouble in breathing might be considered punishment from the spirits of the mountains for his sins. Paracelsus argued instead that miners inhaled metal vapors that damaged their lungs (a condition known today as pneumoconiosis).

In 1527 Paracelsus settled in Basel as the city was turning from Catholicism to Protestantism with riots and calls for revolution. At first Paracelsus was welcomed by the city's conservative leaders. He was appointed the municipal physician and medical professor at the city's university. Students traveled from across Europe to learn from him. But Paracelsus was soon horrifying the authorities by gathering his students to watch him burn Galen's works in public. Martin Luther had stood his ground ten years earlier against Rome, and Paracelsus was doing the same against Galen. As Luther championed the personal experience of the faithful over old rituals, so Paracelsus championed the personal experience of the natural philosopher over the ancient authorities. God had planted his signs in nature for anyone to see without the intervening help of a priest or a physician. Yet even Luther was too conservative for Paracelsus, who declared him no better than the pope.

His iconoclasm got Paracelsus thrown out of Basel, and he spent the last thirteen years of his life wandering, stopping in cities in Bavaria and Bohemia only long enough to alienate the authorities. He became a legend in his own lifetime. Reports would circulate that he had surfaced with some newfound wealth or that he was seen dressed in beggar's rags. By the time Paracelsus died in 1541 at age forty-eight, he had published only a fraction of his writings. The rest he had carelessly left behind on his wanderings. His admirers traveled across Europe to find the lost manuscripts, and, in the decades after his death, slowly brought them into print. Interpreting Paracelsus was no easy task, though. Readers had to pore through dictionaries to understand his impenetrable vocabulary, and faked manuscripts were offered in his name.

The Paracelsists attracted plenty of enemies. The Catholic Church banned his books, while Protestants accused him of witchcraft. One English minister said that Paracelsus was a "drunken conjuror, who had converse with the Devil." To people accustomed to herbs and bloodletting, his metal medicines seemed like poison—John Donne called him the "Prince of Homicide

Physicians." But gradually a Paracelsist movement gained momentum. Paracelsists set out to distill not just the spirit of life, but the spirit of the world itself. Mystical sects like the Rosicrucians celebrated Paracelsus as one of their heroes. Utopian dreamers wanted to turn Bohemia into a Paracelsist country, with Prague as their new Jerusalem, where they would usher the world into a new chemical age. Alchemists, apothecaries, and surgeons took up Paracelsus's medicine. They declared that they could cure the diseases that Galenists dismissed as incurable, such as gout, leprosy, and epilepsy. A French Paracelsist named Theodore de Mayerne became a physician to James I. He organized the *London Pharmacopoeia* in 1618, which he filled with Paracelsus's salts, metals, minerals, and oils. By the early 1600s, Paracelsus was promising to become a real rival to Galen.

The promise didn't last long. The Thirty Years War, from 1618 to 1648, scattered the Paracelsists of central Europe and destroyed their plans for Bohemia. In England, James's toleration gave way to Charles I's conservative hostility. Mayerne lost his place at court, while Laud made Aristotle paramount in the universities and would tolerate no alternatives.

In England, Paracelsus became the hero of the outcast. Many of the utopian victims of the Thirty Years War found refuge in London, where they allied themselves with the persecuted Puritans. The Puritans were suspicious of Galen, finding no worry about sin in his medicine. Paracelsus, by contrast, made medicine a Christian duty. He promised knowledge to anyone willing to read nature's book rather than the books of pagan Greeks. He promised that a prophet-alchemist would arrive soon to liberate the world. Puritans took to Paracelsus immediately, planning Paracelsist colonies in the New World.

The name Paracelsus was never uttered in the lectures that Thomas Willis heard at Oxford. But Paracelsus's cures became popular in England, even if few who used them understood his mystical cosmology. "If I find anything that may be to the good of the patients," said one English surgeon, "be it either in Galen or

Paracelsus, yea, Turk, Jew, or any other infidel, I will not refuse it but be thankful to God for the same." Willis came to appreciate Paracelsus with his bare hands, following his recipes to make drugs for Mrs. Iles. Although Paracelsus had been dead for a century, he reached across the gulf to guide Willis's life. He would help steer Willis toward medicine, and his alchemy would help Willis ultimately find a new account of the soul.

But all that would be twenty years into the future. For now, as the biographer John Aubrey wrote, "his mistress would oftentimes have him to assist her in making of medicines. This did him no hurt, and allured him on."

Chapter Four

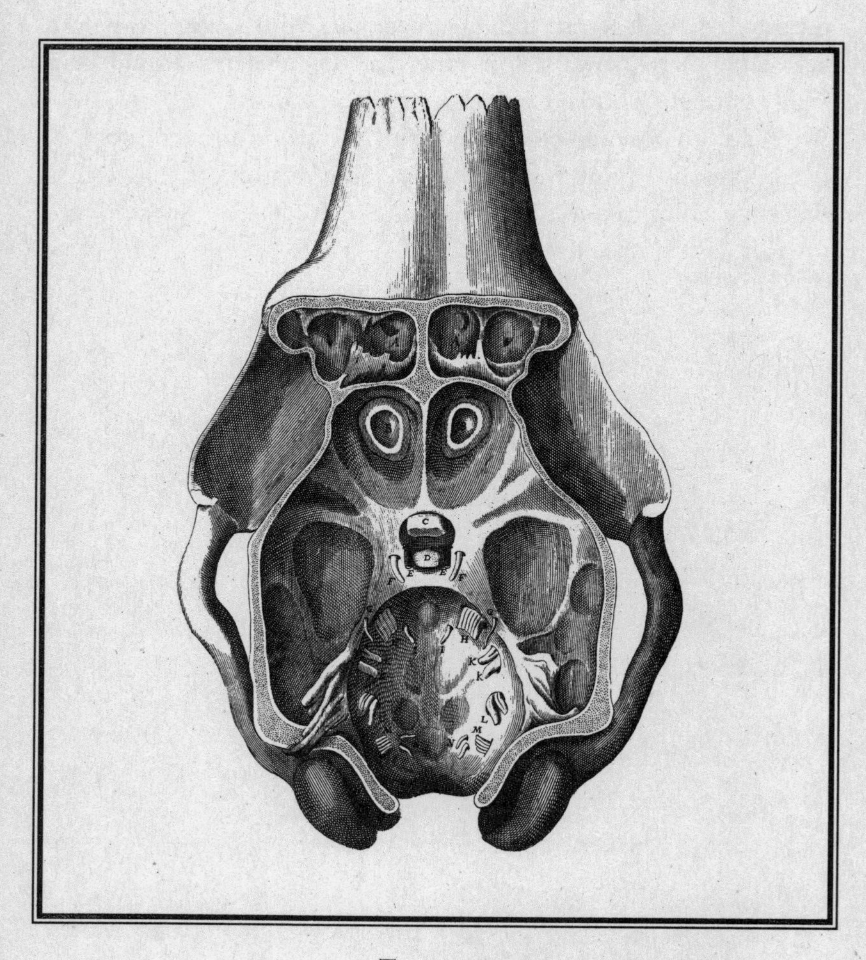

The interior of a dog's skull, from *The Anatomy of the Brain and Nerves.*

The Broken Heart of the Republic

When Thomas Willis became a student at Oxford in 1638, his peaceful world was governed by a divinely appointed king and policed by an all-powerful archbishop. But far beyond his university, events were already unfolding that would eventually devastate his life. His parents would die, his church would be banned, his king would be chased like a criminal, his colleges would be turned into fortresses, and his country would be turned upside down.

Across England, fury was rising against the king. Charles demanded that each coastal county build him a ship to restore the navy, but he happily accepted money instead, and ultimately he demanded the same "ship money" from inland counties as well. As he squeezed money from his subjects, he had Archbishop Laud persecute Puritans and other dissidents in secret courts, where they were sentenced to hideous punishments. In

1637, Charles finally went too far, when he decided that Scotland's churches must use the Book of Common Prayer. Scottish congregations responded to the news with flying footstools and riots. Before long this squabble over church services had evolved into war, with twenty-five thousand Scottish soldiers pouring over England's northern defenses.

At last, after eleven years, Charles was forced to call Parliament again. Expecting to be greeted by docile servants with money at the ready, he faced instead an embittered, strengthened opposition. Angry Puritans and landowners brought up a long list of resentments that had been festering for years. With Scottish forces rampaging through the north, taking Newcastle and its vital supply of coal, Charles had to negotiate with Parliament, agreeing that it would never again be dissolved without its own consent.

Parliament was no longer content with compromise, however. His enemies threw Charles's advisors, including Archbishop Laud, into prison. When Irish Catholics rebelled out of fear that the Protestant Parliament would persecute them, Parliament demanded control of the army in order to crush the uprising. "By God, for not an hour," replied the king.

As Parliament made more demands on Charles—parliamentary approval of his ministers, parliamentary reform of the church—he tried to have his leading opponents arrested. But in a spectacular embarrassment, they slipped away by boat before Charles's soldiers arrived for them. Charles didn't get the news until he came to Parliament to demand the prisoners be brought forward. England's king was humiliated, its Parliament enraged. There seemed no choice but war.

Towns and even families split apart to join either side. Royalists who disliked the way Charles led the country could not bring themselves to abandon their loyalty to king and crown. Many Puritans meanwhile saw the coming war as a stage in a divinely guided history, which would be succeeded by the kingdom of God on Earth. "The turning of a straw would set a whole country in a flame," said one royalist. Charles sent Queen Henrietta Maria to The Hague to sell crown jewels to pay for war, while he left London and headed north to rally royalists to his side.

Traditionally a tranquil town, Oxford turned anxious. Parliament suspected that the university would aid the king however it could, and it was right. The king's supporters smuggled £14,000 in silver plate to him. Hundreds of scholars formed a militia, trading mortarboards for helmets and training for battle in the college courtyards with muskets, pikes, and halberds. Tensions flared between the university and the town fathers, who had always resented the control that the university had over them. The mayor stopped coming to Christ Church to do penance, and the city seized the collection of taxes in the Cornmarket. Antiroyalist mobs gathered in the street to hear the news from London, cheering as they heard of the king's setbacks. Any royalist student passing by risked being assaulted. Puritan students roamed the town, toppling maypoles, which they considered symbols of idolatry.

For Thomas Willis, as for many other young men in his situation, a future as a clergyman must have begun to look bleak. A number of them turned their sights from the church to medicine. Ralph Bathurst, Willis's classmate and close friend, was ordained but soon decided that if he wanted to earn a "tolerable livelihood" he "knew of no way better than to turn my study to physick."

Willis made the same decision. By now, he had already been a student at Oxford for five years, and it would take him another seven years to qualify as a physician. He would have to spend that time listening to lectures on Hippocrates and Galen each week until he was ready to deliver his own series of six one-hour lectures on their work. His official education did not go far beyond the Greeks and Romans; the only anatomy he would learn would come in a single two-day dissection of an executed criminal.

Willis began this obsolete education, learning that the body and temperament were made of four humors, that the animal spirits were created in the marvelous network and carried into the ventricles, that the brain worked like a bellows, pulsating and driving the spirits into the nerves, that you could even see the pulsating brain at work by looking at the quivering skin on a baby's soft head. Willis was taught to resolve any controversy by finding a relevant passage in Galen or Aristotle. But within a few months of starting his medical training, war broke out, and Willis's formal schooling came to a sudden end.

On August 22, 1642, Charles assembled his troops at Nottingham Castle in the English midlands and declared war. Twenty soldiers wrestled the royal standard into the air, its flags flapping in the wind. The highest of the flags read, "Give Caesar his due." Over the next few weeks Charles gathered an army of fourteen thousand men and marched toward London. Parliament had organized its own army, and it blocked his way at Edgehill on October 23. There they fought the first great battle of the English Civil War. The king's army won the day but the victory was Pyrrhic. Its numbers were thinned so badly that Charles realized he could not head for London. Instead, he marched for Oxford, which he turned into a new capital.

When Charles raised his standard, Willis was with his family on their farm outside Oxford. He stayed away from the town for fear that the parliamentary army would force him into service. But the war sought him out anyway. A typhus epidemic broke out twenty miles from his farm in the city of Reading, where the king was fighting Parliament. Years later, Willis would recall the outbreak. "This disease grew so grievous that in short time after, either side left off," he wrote, "and from that time, for many months, fought not with the enemy but with the disease." The king's forces retreated to Oxford, bringing the typhus with them. The soldiers, packed into cramped quarters, began to fall ill. "The very air seemed infected," Willis wrote.

Willis was still only twenty-one, with only a few months' medical education, but he had already developed a doctor's eye. He observed the disease as it spread from the returned soldiers to the rest of the army. The fever would not have a name for centuries, but Willis's careful observations of the epidemic represent the first clinical description of typhus. It soon spread from the soldiers to the town and countryside. "Here this disease soon became so epidemical that a great part of the people was killed by it," Willis later wrote, "and as soon as it entered a house, it ran through the same, that there was scarce one left well to minister to the sick."

As the epidemic spread, Willis kept track of the whelks and spots that covered the skin of the fever's victims, the disordered pulse and raging heat. And he was also struck by the way in which the disease altered not only its victims' bodies but their spirits. Some people fell into stupefaction, some people began to tremble and convulse, others spun with vertigo.

Willis recorded the epidemic with clinical detachment, even as it swept away the last vestiges of his own childhood. His father, who had joined the royalist army to defend his king and church, died of the camp fever, and ten days later his stepmother also died. Willis, the eldest son, now had to take his father's place. He had to raise his six brothers and sisters and run their 100-acre farm. The job would have been hard enough in peacetime, but as the Civil War spread, it became impossible. Parliamentary soldiers seized the town of Abingdon near Willis's farm and rode out into the surrounding countryside to steal cattle, horses, and food. Although the king was only two miles away in Oxford, his soldiers couldn't protect people in the countryside such as Willis. The meadows through which he had walked to school as a boy became a war zone.

In later years Willis would look back on the epidemic that killed his father and stepmother as emblematic of England's disorder. The magistrates were so preoccupied with war that they allowed people to suffer in squalor, giving diseases the chance to spread. Only if order was restored, he was convinced, would such outbreaks subside. Willis was not alone in his belief. In 1643 a royalist pamphlet spread through London, presenting itself as a plea from the city's residents to the king. It begged Charles to return from Oxford so he could cure Londoners of the King's Evil.

Charles's return would also bring a cure to England itself, "which hath languished of a tedious sickness since your Highness departure from thence, and can no more be cured of its infirmity than we, til your gracious return hither."

—

Parliament was advancing on the king across the entire country, and at the end of 1643, it made an alliance with the Scots, who sent twenty-one thousand soldiers marching south. The only advantage Charles had left was his well-trained, disciplined army. In April 1644, Charles called for loyal citizens around Oxford to come defend the city as volunteers, so that he could send his own soldiers into battle in other parts of the country. Thomas Willis decided to leave his brothers and sisters and make his way back to Oxford to become a soldier.

Willis had been away from Oxford for less than two years, but by the time he returned, the city had been drastically transformed. Charles had turned it into a fortress, with dams to flood the surrounding meadows and a giant earthen wall along its northern boundary, all built by the town on his orders. The university, Willis found, had suffered for its loyalty. Charles had "borrowed" thousands of pounds to pay his army, and most students had left to fight for the king or to escape the war. The few students remaining at the university were pushed into a "dark, nasty room" in New College. The king seized Christ Church for his own quarters, where he could sometimes be seen walking his dogs across the courtyard. The rest of the university was taken over by the army. The school of law was filled with supplies of corn and cheese, the school of astronomy with soldiers' uniforms, the school of rhetoric with rope bridges. Cannon were hauled up into the highest student rooms, and the lead roofs of the chapels were torn off and melted down for bullets.

The city had swelled like a boil. Seven thousand soldiers were now stationed there, with a gallows built in the middle of town for military executions. Charles had set up his own parliament and law courts in Oxford, along with a mint to turn its silver plate into coins. The royal court squeezed in as well, bringing children, mistresses, and pastry cooks. Royalist writers set up shop, cranking out pamphlets and satires attacking Parliament's army as buffoons and atheists who preferred sex with horses to women. Londoners driven out by the Puritans—artists, musicians, astrologers, actors—found shelter in Oxford wherever they could. Even cattle were brought within the city's walls; royalist soldiers stole herds from the surrounding countryside and corralled them in Christ Church, where their owners could come the next day and buy them back. Spies from Parliament infiltrated the city, along with prostitutes believed to have been sent from London to spread syphilis through the king's army.

Oxford had had a reputation as a healthy country city, but now it degenerated into a sick, foul slum. Not long after Willis took up arms, a fire broke out when a soldier tried to roast a stolen pig. It swept from the north wall of the city southward over the course of ten hours, feeding on the stacks of furze, corn, and hay. With Parliament's soldiers watching the smoke rise over the surrounding hills, the royalist army could not leave the walls to help put the fire out. A sixth of the city burned down, leaving hundreds of poor families to wander the streets or huddle in the houses of friends.

Rats multiplied and spread plague through Oxford, and with only a few doctors willing to treat its victims, another thousand lives were lost in 1645 and 1646. Corpses were left lying in the streets for days before they were buried. The poet John Taylor described the Thames flowing through Oxford at the time:

Dead hogs, dogs, cats, and well-flayed carrion horses
Their noisome corpses soiled the water sources;
Both swines' and stable dung, beasts' guts and garbage
Street dirt, with gardeners' weeds and rotten herbage.
And from this water's filthy putrefaction, our meat and
drink were made, which bred infection.

At the university, Thomas Willis trained alongside other young royalists whose path to the church had been blocked by the war. Stationed along the northern walls of the town, they faced parliamentary forces that ringed the hills. From time to time, they fought in skirmishes and raids, but for most of the war they guarded the walls. The king's artillery, positioned behind Willis in the grove of Magdalen College, fired cannonballs over his head. When his company was not firing at the enemy, they stood on the walls smoking clay pipes or drinking sour beer.

The diarist Anthony Wood lived in Oxford during the war and watched these volunteers lose their innocence. "As for the young men of the city and university he found many of them to have been debauch'd by bearing armes and doing the duties belonging to soldiers, as watching, warding, and sitting in tipling-houses for whole nights together."

But Willis put the time to good use by teaching himself medicine. The old Galenist traditions at Oxford had fallen apart—not because Paracelsists had stormed in to set the old books on fire, but simply because the war had devastated the university. No lectures were held, no cadavers dissected. Willis got hold of whatever books he could and talked to anyone he could find with even a smattering of medical knowledge. Although Willis respected Galen—could fourteen hundred years of tradition be altogether wrong?—he was open to new ideas, no matter how disreputable their associations.

Willis was not the only young man in Oxford at the time interested in medicine and the workings of the body. A few curious scholars still lingered in town, and they had discovered that one of the king's physicians was a teacher without equal: a bitter old man named William Harvey.

—

William Harvey, physician extraordinary to Charles I, moved like a pale vision through the Oxford nights. His hair and pointed beard were as white as bleached bone. His small body moved through the streets with a stiffness brought on by years of gout. His eyes were sharp and dark, like an owl's, yet they seemed to be turned inward. "He did delight to be in the darke," his friend John Aubrey would later write, "and told he could then best contemplate."

Twenty years earlier, Harvey had made a discovery so simple that it could be summed up in a few words: the heart sends blood through the body in a loop. Harvey did not make this discovery as a revolutionary. He did not proclaim, as Descartes would have done, that his discovery was the fruit of a new kind of science. Harvey scoffed at Descartes and the rest of the scientific revolution. His hero was Aristotle, not Galileo. Yet today we can see that Harvey was to medicine what Galileo was to physics. He reached down into the very core of Galen's account of how the body worked and showed that it was flawed. As important as Harvey's discovery of the circulation of the blood was, however, it was eclipsed by another of his discoveries: a new way to study the body. Like Vesalius, Harvey learned to look at the body for himself, on top of which he tested his observations with experiments.

Four decades earlier, in 1600, Harvey had begun his medical career in Padua. As a medical student, he went to the university's anatomy theater, a windowless colosseum made out of concentric rings of balconies packed with hundreds of physicians, students, and gawkers. Lit by candles at the bottom of the theater's well, an anatomist opened a criminal's corpse, as Vesalius had, with his own hand, and described to the audience overhead the arrangements and functions of every muscle, of every bone, until all that was left was fit only for dogs to eat.

At Padua, Harvey became steeped in Aristotle. Six out of the eighteen professors lectured on Aristotle and nothing else. The anatomists at Padua studied the body as Aristotle had, examining each organ to understand how it served the soul's fulfillment. To better understand an organ's purpose, Aristotle had compared the same one in different animals, and the Padua anatomists did the same. Reviving this ancient method, they found new parts of the body and came to a better understanding of how the entire body works.

Harvey studied under Hieronymus Fabricius, who, among other things, showed that the eyes' pupils respond to light and gave the first proper account of the placenta's function. In his work, Harvey later said, he followed Aristotle as a leader and Fabricius as a pathfinder. Most of his students couldn't understand why Fabricius wasted so much time studying and lecturing about animals, but Harvey did. He also understood that Fabricius made a peculiar mistake about the anatomy of the veins.

Fabricius explained that the veins in many animals, including humans, have valves—little doors, as he called them—situated at the points where the smaller vessels branch from the main ones. Fabricius tried to figure out what purpose they served. After comparing the little doors in various animals, Fabricius decided that they must slow down the blood as it flowed away from the heart through the veins, ensuring that the body didn't get overwhelmed with blood. For all his skill, Fabricius was wrong. The little doors actually opened up toward the heart, not away from it.

At some point, Harvey noticed this mistake, either while he was still in Padua or after he returned to England, in 1602. As he built a prosperous London medical practice, he carried on the anatomical research in which Fabricius had trained him. He dissected chicks and dogs, frogs and eels, fish and pigeons. Harvey's chief interest was in the blood, the heart, the veins, and the arteries. He pushed probes into the veins of cadavers and found that they slid easily through the little doors if he pointed them toward the heart. When he tried the other way, he could not get through. The doors sealed themselves like a floodgate.

Harvey recognized that the little doors had a huge importance. If they were designed for blocking the flow of blood, that meant that the veins must carry blood toward the heart, not away from it, as Galen had taught. God would not create different parts of the body at cross purposes. Yet if the blood of the veins flows toward the heart, Galen's entire account of how the body works was called into question. To get to the bottom of the matter, Harvey did what Aristotle would have done. He looked at animals to find what was common to their hearts and blood vessels. He cut open fish, for example, to watch their slow-beating hearts, each beat distinct from the next. He saw the blood from the veins flow into the heart and then into the arteries. In other words, there was no difference between the blood in the two sets of vessels.

The more Harvey studied the blood, the more vital he found it to be. Blood was not just another humor but the primary substance of the body. When he cracked open chicken eggs, he could see blood moving through the embryos at even the earliest stages, long before the organs of the body had formed. This sequence made no sense if, as Galen claimed, the liver manufactured blood. "Blood is rather the author of the viscera than they of it," Harvey decided. He might be a respectable, conservative doctor, but he was, without entirely realizing it, becoming deeply radical.

Harvey was so respectable that in 1616, at age 38, he was chosen by the Royal College of Physicians to teach anatomy to London's surgeons. Surgeons were a rung down the professional ladder from physicians in the 1600s, amputating legs, putting leeches to patients, and doing the rest of the manual labor of medicine. But the College of Physicians offered surgeons some academic training and even a little philosophy to underpin their work. Twice a week through the year, Harvey would lecture to them standing before a cadaver, dressed in a puffy bonnet and an apron covering him from shoulder to knees, using whalebone rods tipped with silver to point out tendons and joints.

In his lectures, Harvey offered no great earthshaking discoveries. Still, for an English physician in the early 1600s, he was peculiar. Instead of presenting Galen to his students, he told them about experiments he had carried out—how, for example, he had slit the heart of a pigeon and watched the blood spurt; how an hour later its heart was still beating. The cadavers Harvey dissected might be those of criminals, but his own body and those of his family fascinated him just as much. He described his sister's five-pound spleen, and he recounted how his own gold ring had passed unharmed through his bowels. Harvey trusted his own observations even when they didn't conform with Galen's (and despite the fact that any public disagreements with Galen could draw a fine). Galen claimed that human lungs have five lobes, but Harvey told his surgeons that they in fact had four. He tried to make his defiance as diplomatic as possible. "Perhaps in Galen's time it was common in man, whereas now it is rare," he said.

Harvey laid out his lectures like a meal, with courses of viscera, muscles, and bones. He ended them with what he called "the divine banquet of the brain." But this last course was more like a dish of jumbled leftovers. The head is round, Harvey explained in good Aristotelian fashion, because "roundness is the most perfect shape." On other matters, he deferred to Galen instead of Aristotle. In Harvey's day many of Aristotle's followers still believed that the brain was insignificant. "The heart is not only the origin of all the veins but also of the nerves," declared the Italian anatomist Andrea Cesalpino in 1588. Harvey granted that Galen was right on this score. "The brain itself neither sees nor hears, and so forth, yet it does all these things," he declared.

Harvey had no new ideas for how the brain could do these things. He mentioned ideas about how the furrowed convolutions on the surfaces might help the brain pulsate or keep the ventricles from collapsing shut. Harvey still went along with Aristotle on the coldness of the brain. It is cold, he explained, "in order to temper the spirit coming from the heart, lest it be overheated and quickly evanesce, as happens in madmen whose brains are heated." The rest of his lecture was a grab bag of old notions—that the brain increased in size with each full moon, for example, and that hysteria originated in the womb.

Harvey was certainly interested in the brain, and he was even something of a proto-pyschologist. He had a reputation as one of the few physicians in London who could cure devils in the head. But he left behind no theory of his own as to how the brain worked. His silence may have come from the fact that Harvey did not believe that the brain was all that important—that it was not, as he put it, "the first principle" of the body. Long before an embryo developed a brain, it was sensitive. "If this first rudiment of the body be gently pricked, it will, like a worm or grub, obscurely move and contract and twist itself, which is plain evidence that it has sensation," he wrote. "It is clear that all sensation and movement does not derive from the brain." The blood could feel, Harvey concluded. "That sensation as well as movement is inherent in the blood is obvious," he wrote. The soul itself—at least the vegetative and sensitive souls—could be found in the blood, Harvey believed, just as the Bible taught.

Given all the powers of blood, Harvey saw no need to bother with spirits. Galen's followers believed that the spirits of the liver, heart, and head were all essential to the body's survival, and they had been joined in the 1500s by physicians inspired by Plato, who held that the world-spirit infused the body with spirits of its own that made life possible and was the secret of health. "I have never been able to find any such a spirit anywhere," Harvey complained. He could not even find a place where they could be generated. "I hold that the ventricles of the brain are not suited to this so excellent function, for I believe them rather to be made for the reception of excrements."

Blood, Harvey believed, did everything that the spirits were given credit for. According to Harvey, spirits did nothing more than "serve as a common subterfuge of ignorance. For smatterers, not knowing what causes to assign to a happening, promptly say that the spirits are responsible (thus introducing them upon every occasion). And like bad poets, they call this deus ex machina on to their stage to explain their plot and catastrophe."

—

Harvey himself joined the cast of an extravagant play in 1618, when he took on the minor role of physician extraordinary to the royal court of James I. He had a few speaking lines over the next three decades, mostly at births and deathbeds, but otherwise he stood to one side along with the rest of the court, witnessing an unfolding tragedy. During his time at court, he served his king with deep loyalty, writing how "the heart of animals is the foundation of their lives, the sovereign of everything within them, the sun of their microcosm, that upon which all growth depends, from which all power proceeds. The King in like manner is the foundation of his Kingdom, the sun of the world around him, the heart of the republic, the foundation whence all power, all grace, doth flow."

When Harvey was not treating the king or his court, he went on investigating the heart and blood. Galen had argued that blood flows through two separate systems of veins and arteries. As the body devoured the blood, it had to be replenished by food, just as rivers were refreshed by the snow falling on mountains. Harvey's own experiments left him disenchanted with Galen's account, and he now looked for a different one. Trying to explain the flow of the blood, Harvey turned back, once again, to Aristotle. "I began to bethink myself whether it had a sort of motion as in a circle," Harvey wrote.

A circle appealed to Harvey, since Aristotle taught that it was geometry's perfect shape. Harvey envisioned blood traveling through the veins toward—not away from—the heart. Now the little doors discovered by Fabricius made perfect sense. When blood from the veins entered the heart, it then traveled to the lungs and back to the other side of the heart. From there it entered the arteries, traveling through them to the ends of the body. The arteries were linked to the veins by hidden passageways so that the blood could pass from the former to the latter. Once the blood entered the veins, it came back toward the heart. Harvey realized that the pulse was caused when the heart drove surging waves of blood into the arteries. When Harvey pierced the hearts of pigeons, it was the red blood of arteries that came squirting out, not the purplish blood of veins.

Reordering the body this way was drastic enough, but Harvey then did something unheard of in Europe. He carried out experiments to test his theory. Galen claimed that blood from the veins trickled through invisible pores in a wall inside the heart to get into the arteries. Harvey injected water into a chamber of an ox heart and saw that none of it could cross the wall. "But by Hercules! No such pores can be demonstrated, nor in fact do any such exist," he wrote.

Harvey estimated how much blood flowed out of the heart. He calculated that each beat drove half an ounce of blood into the aorta. In half an hour, the heart would release three pounds of blood. But when Harvey drained all the blood from a sheep, he found that its entire body contained only four pounds. Galen had claimed that the body simply used up the blood supplied to it by the veins and arteries. If he was right, every animal would have to be an outrageous glutton just to replace the blood that left its heart. Harvey faced no such problem with his own theory. The heart could recycle the same few pounds of blood over and over again.

To see whether the blood actually did flow from the arteries into the veins, Harvey tied off the arm of a man. The arteries just upstream from the ligature swelled up, and when he untied it, the man's hand turned dark and warm and swollen as the blood surged in. When the ligature was tied loosely, the heart could drive the blood into the arm through the arteries, but the veins could not push back their weaker flow. The man's arm began to swell.

Harvey then tied up a man's arm once more to investigate how the little doors worked in the veins. Once the blood stopped flowing into the arm, Harvey pressed down on a vein with a finger. With another finger he stroked the vein's blood up the arm, past one of the little doors. He could see that segment of the vein empty out and remain empty, just as he predicted. In experiment after experiment Harvey saw his theory confirmed.

He also knew that his experiments were throwing the entire system of European medicine into question and raising the possibility of creating a new one to take its place. "My whole life, perchance, would not suffice for its completion," he wrote.

But Harvey had no new medicine to offer King James when his health began to fail. When Charles inherited the throne, Harvey continued to serve as a court physician, and the two men came to be friends. The king hunted deer in the royal parks almost every week and allowed Harvey to dissect his kills. Harvey was most interested in pregnant does, which he opened up to search for embryos. Charles was, Harvey wrote, "himself much delighted in this kind of curiosity and was pleased many times to be an eyewitness to my discoveries." One November day, Harvey removed a minuscule embryo from a doe, with blood already pulsing through its vessels. He showed it to the king. "It was then so small," Harvey later wrote, "that, without the advantage of the sun's beams obliquely falling upon it, he could not have perceived its shivering motion."

At other times, Harvey showed Charles how the blood circulated. His king, it turned out, was one of his few sympathetic listeners. In the mid-1620s Harvey explained his theory to the College of Physicians, performing his demonstrations to persuade them. He then published a slender book in 1628 called *De motu cordis (The Motion of the Heart),* in which he presented his argument with terse, elegant precision. He dedicated it to the College. "I was greatly afraid to suffer this little book, otherwise perfect some years ago, either to come abroad, or to go beyond the sea," he wrote, "lest it might seem an action too full of arrogancy, if I had not first propounded it to you, confirm'd it by ocular testimony, answer'd your doubts and objections, and gotten the President's verdict in my favour."

His efforts to woo the College were wasted. His fellow physicians generally dismissed the book. The physician James Primerose fired off the first attack in 1630, which consisted of little more than quotations from Galen and a few other authorities. Since Harvey disagreed with them, Harvey must be wrong. End of argument. Other physicians found it absurd that anyone would make such enormous claims based on autopsies. Death, they believed, disturbed the body beyond repair by stirring up the four humors and making the pores of the heart collapse.

To many physicians, Harvey simply did not speak the language of medicine. They opened his book expecting a formal philosophical argument. Instead, they found something else entirely. "I profess to learn and teach anatomy not from books but from dissections," Harvey declared, "not from the tenets of philosophers but from the fabric of nature." To his contemporaries, Harvey seemed like the body's accountant, tallying up ounces and quarts when he ought to be thinking like a philosopher.

A few critics did ask some good questions. Harvey claimed that the blood returned from the arteries to the veins, but he failed to show where this happened. And what was the purpose of the circulation? A reader finished *De motu cordis* with a vision of blood whirling pointlessly in a circle. Most important of all, why should a physician care? Harvey did not show how medicine should change in the face of his discovery.

For years after Harvey published *De motu cordis,* hardly anyone in England accepted his idea. Outside England, his reception was a little more positive. Descartes praised Harvey in his *Discourse on Method* in 1637 and claimed that the circulation of the blood made perfect sense in a mechanical body. But Harvey didn't show much interest in the news. He considered Descartes a careless anatomist and had no patience for newfangled accounts of the body as a machine—a view which at the time was still limited to a few mavericks. To Harvey, it was absurd to think of the body as a soulless contraption built from lifeless particles.

De motu cordis would become a landmark in the history of science. But to Harvey it felt more like a millstone than a milestone. His friend John Aubrey wrote that after the book was published, Harvey "fell mightily in his practice, and that 'twas believed by the vulgar that he was crack-brained; and all the physicians were against his opinion and envyed him." Harvey bitterly watched his reputation suffer. "It cannot be helped that dogs bark and vomit their foul stomachs," he groused, "but care can be taken that they do not bite or inoculate their mad humours, or with their dogs' teeth gnaw the bones and foundations of truth." As angry as his critics made him, Harvey chose not to answer them for twenty years. "I think it is a thing unworthy of a philosopher and a searcher of the truth, to return bad words for bad words," he wrote.

Instead, Harvey went back to his work. He studied embryos and made a massive study of insects, comparing them to animals with backbones. Harvey and his king, two prickly, philosophical men who felt abused by the world, were drawn into close friendship. They would even give each other gifts—a pound and a half of silver plate from the king, a box of marmalade from Harvey in return.

Charles helped Harvey in his research whenever he could. In 1641, the king sent him to meet the viscount Montgomery, a man rumored to have an open chest. Charles had heard that Montgomery had been wounded as a boy in a fall from a horse, the injury leaving him with a gaping hole ever since. He asked Harvey to investigate. When Harvey met Montgomery, he was astonished that the young man was in good health. Montgomery astonished him more by showing him a metal plate on his left side, which he then pulled away. "I immediately saw a vast hole," Harvey later wrote. Through it, Harvey could see for the first time a beating human heart. Before that moment, Harvey wrote, "I was almost tempted to think . . . that the motion of the heart was only to be comprehended by God."

The hole in Montgomery's chest was big enough for Harvey to stick in three fingers and a thumb. He could feel the systole and diastole, the blood surging into the heart and then flushing out, exactly in the order he had predicted. Pressing a finger to Montgomery's wrist, Harvey could feel the pulse arriving as he proposed, a moment after the heartbeat. "Instead of an account of the business," Harvey wrote, "I brought the young gentleman himself to our late King, that he might see, and handle this strange and singular accident with his own senses."

Charles reached into the cavity and felt the recoiling, rushing heart. "Sir," he said to Montgomery, "I wish I could perceive the thoughts of some of my nobilities' hearts as I have seen yours."

The king never figured out how to read the hearts of his subjects and instead ended up in a civil war. Harvey followed Charles to Nottingham when he raised his standard, and at the battle of Edgehill he protected the king's sons under a hedgerow. Finally, Harvey came to Oxford with the rest of the court, where he became virtually a prisoner. At Whitehall he had left behind his wife and an apartment packed with medical papers—nearly forty years of observations of his patients, notes on postmortems, studies on animals of all sorts with special attention to insects. "He had studied the sea and land, islands, and continents, mountains and valleys, woods and plains, rivers and lakes, and whatever mysteries they contained," a friend later wrote. "He pursued most diligently the origin of all creatures, their food and habits," and had written about the brain and "the movement and feeling of animals."

At its very outset, the war delivered Harvey a wound from which he never recovered. "While I did attend upon his most Serene Majesty in these late distractions and more than Civil Wars," he wrote later, "not only by the Parliament's permission but by its command, some rapacious hands spoiled me of all the goods in my house and (which I most lament) my adversaries stole from my study the notes which cost me many years industry."

Everything was lost when looters stormed Harvey's abandoned rooms at Whitehall. It was a final sour proof of something Harvey had suspected for years. "Man," he declared, "is but a great mischievous baboon."

When Thomas Willis volunteered as a soldier in 1644, Harvey

had been trapped in Oxford for over a year. He treated sick members of Charles's court for plague, typhus, and malaria, although in later years when he spoke about the siege he said that "he observed more people die of grief of mind than of any other disease." Harvey's own sadness increased as well. His wife died in London, and Parliament, not satisfied with having destroyed his life's work, had him fired from his post at St. Bartholomew's Hospital.

But Harvey brightened when he worked with the young men at Oxford. Prior to the war, only one English physician had gone on record in support of his theory of circulation. But when Harvey showed Oxford students how the blood circulated in the body, many of them quickly accepted it. Even more important, they accepted his methods. Every day Harvey would visit George Bathurst (the elder brother of Ralph, Willis's close friend and fellow frustrated clergyman). George kept chickens in his room, and they would crack open the eggs and study the embryos. As they looked closely at chickens, Harvey explained his emerging philosophy about the dawn of life. Embryos do not start out with all their parts fully formed. Instead, every animal begins as an egg that has its own vital powers—a soul, which develops new organs gradually.

The experience of working with the old doctor was to leave its mark on the young men for the rest of their lives. After the war, Harvey's Oxford disciples would dedicate themselves to learning about the body through observation and experiment. Thomas Willis would spend the rest of his life following Harvey's example. After the war, he would search for the true nature of blood, and in later years the blood would lead him to the brain, the nerves, and ultimately the soul.

—

If Thomas Willis had borrowed a spyglass as he stood on the Oxford walls in 1645 and scanned the surrounding hills, he might have caught sight of a man who would soon rule over him and the rest of England. A heavy-browed colonel named Oliver Cromwell arrived that year to help the parliamentary general Thomas Fairfax tighten the noose around Oxford.

Cromwell didn't have a day of military experience before the war, but in battle he proved to be an extraordinary leader. By 1643 Cromwell had become so successful that Parliament gave him permission to reorganize its army, turning a messy rabble into a well-disciplined organization joined by an equal duty to God. The New Model Army, as it became known, was in many ways a realization of the Puritan ideal: an efficient army less interested in plunder than in preparing for the return of Jesus Christ.

Although Cromwell was convinced he was fighting God's war, he did not become a religious tyrant. If a soldier accepted the basics of Protestant Christianity, he was free to hold whatever other opinions he wanted. "I had rather that Mahometanism were permitted amongst us, than that one of God's children should be persecuted," he later said. Although Cromwell didn't realize it at the time, he was also setting up a new experiment in democracy. Nothing like it had been seen before. Men of ordinary means became part of a miniature society in which they could openly debate the most shocking, radical ideas and print pamphlets for their fellow soldiers to read.

The army's debates revealed just how many different factions opposed the king. Presbyterians, who dominated Parliament at the time, claimed that the authority of the church came from God, not man. They wanted the clergy to be invested with divine authority and laymen ordained with a laying on of hands. Cromwell belonged to the Independents, who wanted each congregation to govern itself. Political movements sprang up in the New Model Army as well. The Levellers, for example, wanted to return England to their notion of political life before the Norman conquest: an Anglo-Saxon utopia in which all householders would have the vote and the monarchy would be abolished—"every man by nature being a King, Priest and Prophet in his own natural circuit and compass," as one Leveller wrote.

Voices rose from the army calling not only for political democracy but also for a kind of religious democracy as well. Sects emerged among the soldiers that rejected any authority beyond the individual. In January 1646, Presbyterian ministers traveled to the outskirts of Oxford to visit Cromwell's soldiers, where they got an unwelcome surprise. As they later reported, "The multitude of soldiers in a violent manner called upon us to prove our calling . . . whether those that are called ministers had any more authority to preach in public than private Christians which were gifted." Army chaplains urged the soldiers outside Oxford "to turn the world upside down."

Ideas once unmentionable now raced across England. The Leveller writer Richard Overton dared to deny the existence of the immaterial soul in an outrageous pamphlet called *Man Wholly Mortal*. If we reasoned with a soul distinct from our bodies, Overton asked, did that mean that drunks lose their souls after too much beer? Where was eternal reason in a child? "Rationality in an infant," Overton declared, "is no more in it than a chicken is in an egg."

Overton mocked the Aristotelian idea that the soul had many different faculties, each of which was presumably immaterial as well—"a rational soul, a memorative soul, a seeing soul, a hearing soul, a smelling soul, a tasting soul, a touching soul, with divers other souls of all sorts, and sizes: as, saving your presence, an evacuating soul, &c." These so-called faculties, Overton declared, were nothing more than physical features of the human body itself. "Man is but a creature whose several parts and members are endowed with proper natures or faculties each subservient to each other, to make him a living Rational creature."

Overton's was far from a lone voice. The idea that the soul was material and died with the body—known as mortalism—had been popular for decades both in remote villages and in the slums of London. Now that the Civil War had brought England to a boil, slum religion could get the attention of the whole country. Even a few university-trained writers such as John Milton believed that the soul was mortal. Mortalists denied they were heretics, claiming instead they embraced the true Christianity. If the soul were immaterial, it would have no way to break out of a material body at death. "He so martyred has an ill-favored Paradice for his soul," Overton wrote. Since Adam's fall was of both body and soul, Overton argued that his soul could not have possibly remained immortal. The faculties of the fallen died with them, just as those of animals did. "The invention of the soul upon that ground vanisheth," Overton concluded.

To royalists and loyal followers of the Church of England, mortalists represented one of the gravest threats to the nation's salvation. The pamphleteer Alexander Ross called Overton "a pig from Epicurus's stye." From Oxford under siege, the royalist Guy Holland accused Overton of "depressing man even as low as brute beasts, and ascribing to them both a mortality alike." The only reason Holland could imagine that Overton would write such lies was because he wanted to be excused for indulging his lusts during this life.

Through 1645, these mortalists, these Levellers, these many threats to the conservative order pressed toward Oxford's walls. In London, meanwhile, Parliament beheaded Archbishop Laud for treason. Across the kingdom, Cromwell's New Model Army proved its effectiveness by defeating Charles's forces in battle after battle. Puritans swept into captured towns and ransacked the cathedrals. The royalist outposts around Oxford began to fall, and by the end of 1645, the king was ordering his sons to slip out of Oxford. They resisted at first, but finally they left their father and made their way to Paris, where they joined their mother, Queen Henrietta Maria. In April 1646, the king cropped his hair, shaved his beard, disguised himself as the servant of an army paymaster, and slipped out of Oxford.

Charles left his army and court to the mercies of the New Model Army. General Fairfax, who had been a student at Oxford himself, respected the university, but knew that his army might not. To many Puritans, Oxford represented everything they most despised about England's traditional order, everything that needed to be reformed or even destroyed. They referred to the colleges as "the nurseries of wickedness" and "cages of unclean birds." Radical ministers declared that the Bible had prophesied the destruction of the universities so that the Lord could teach his people directly. Fairfax negotiated with the abandoned royalists, writing, "I very much desire the preservation of that place (so famous for learning), from ruin, which inevitably is like to fall upon you, except you concur." Their response was a blast of artillery fire.

By June 1646, the royalist cause could not hold out any longer. The Oxford forces negotiated a surrender that allowed them to leave the city unmolested, although many a man among them was given six months to get his affairs in order and "to go to any convenient port, and transport himself with his servants, goods and necessaries beyond the Seas." Carriages of weeping noblewomen rolled over Magdalen bridge, followed by the defeated soldiers, marching out of the city past the parliamentary lines. Thomas Willis marched out with them, the world of his childhood shattered and his future unsure. Harvey slipped out of Oxford as well and into obscurity.

The defeated army dissolved away. The leading royalists joined the king's family in Europe, while many less fortunate soldiers were captured and shipped to the new colonies in Barbados to work like slaves. Some soldiers managed to return to their former homes and former lives, while others retreated into the forests to rob travelers.

Whatever their fates, these royalists all shared in the devastation of their king's defeat. They emerged from Oxford's walls into a new world in which, according to the poet and royalist soldier Richard Lovelace, "the dragon hath vanquished St. George." With the king's defeat, Lovelace wrote,

Now the Sun is unarmed
And the moon lies as charmed
And the stars dissolved to a jelly,
Now the thighs of the Crown
And the arms are lopped down,
And the body is all but a belly.

Chapter Five

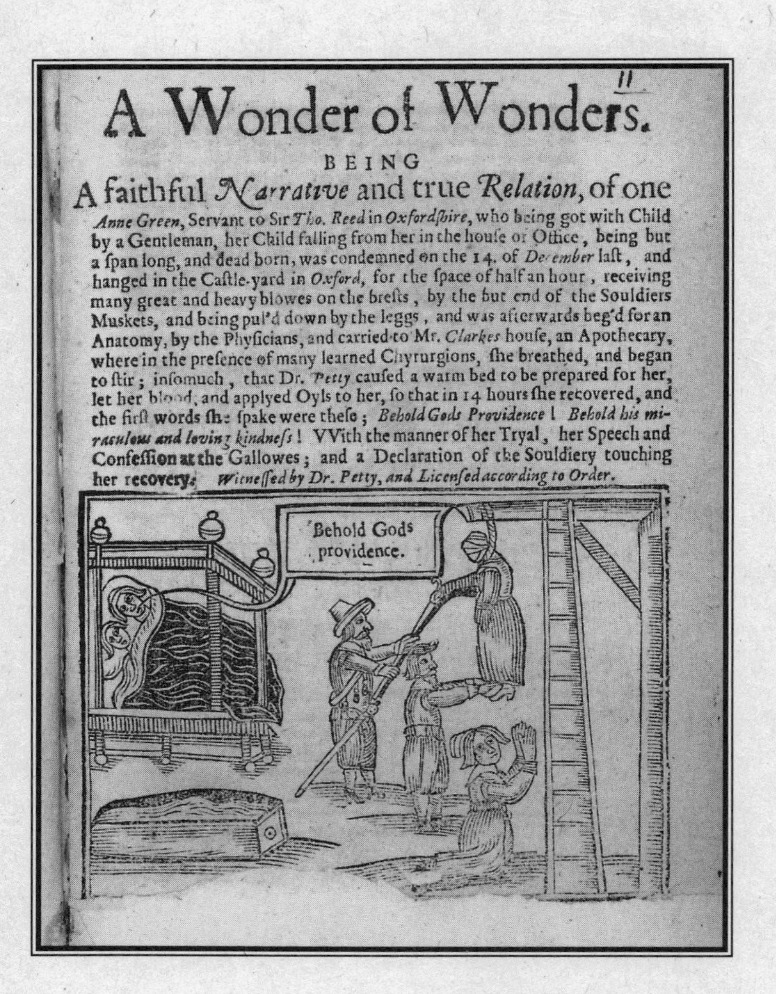

A Wonder of Wonders.

BEING

A faithful *Narrative* and true *Relation*, of one *Anne Green*, Servant to Sir *Tho. Reed* in *Oxfordshire*, who being got with Child by a Gentleman, her Child falling from her in the house of Office, being but a span long, and dead born, was condemned on the 14. of *December* last, and hanged in the Castle-yard in *Oxford*, for the space of half an hour, receiving many great and heavy blowes on the brests, by the but end of the Souldiers Muskets, and being pul'd down by the leggs, and was afterwards beg'd for an Anatomy, by the Physicians, and carried to Mr. *Clarkes* house, an Apothecary, where in the presence of many learned Chyrurgions, she breathed, and began to stir; insomuch, that Dr. *Petty* caused a warm bed to be prepared for her, let her blood, and applyed Oyls to her, so that in 14 hours she recovered, and the first words she spake were these; *Behold Gods Providence! Behold his miraculous and loving kindness!* VVith the manner of her Tryal, her Speech and Confession at the Gallowes; and a Declaration of the Souldiery touching her recovery. *Witnessed by Dr. Petty, and Licensed according to Order.*

THE TITLE PAGE OF A PAMPHLET DESCRIBING THE MIRACULOUS RESURRECTION OF ANNE GREEN AT THE HANDS OF THOMAS WILLIS AND WILLIAM PETTY.

Pisse-Prophets Among the Puritans

For three months after the fall of Oxford, Thomas Willis disappeared into one of history's fog banks. Wherever he went, by September 1646 he was back in Oxford. His king was in flight and his army destroyed, but Willis was greeted as a returning hero.

When Parliament's soldiers had entered the university, they hadn't done much damage beyond throwing a few surplices down a latrine, for General Fairfax had made good on his promise and posted guards at the doors of the Bodleian Library to protect it from looters. The Puritans were more interested in bringing the corrupt souls of Oxford to God. A flood of haranguing ministers—some of them onetime Oxford students who had been thrown out by Archbishop Laud—streamed into the city. Games were banned; holly was forbidden at Christmas; the colleges and chapels were stripped of their idolatrous images of Jesus, the Virgin

Mary, and the saints. Soldiers strode into lectures to yell at scholars for indulging in human learning, challenging them to prove their calling from Christ. Parliament sent seven Presbyterian ministers from London to set the university on the proper path. They halted the Latin masses and removed the popish altars and paintings from the chapels. They did not get a hearty welcome. The diarist Anthony Wood called them knaves, fools, and madmen.

But Parliament did not deal with the university more thoroughly, because it was distracted by much greater concerns. Without enemies to unite them in war, the Presbyterian and Independent factions in Parliament now fell out with each other. They debated what, if anything, should replace the old Church of England. Soldiers in the New Model Army were furious that they were not being paid for their services to England and, thanks to Cromwell, were organized well enough now to protest. The more radical Levellers in the army were even agitating to make England into a kingless, democratic country.

Parliament also still worried about Charles I. After fleeing Oxford, the king had reached the Scottish forces in the north of England, hoping to persuade them to side with him against Parliament. For months, the Scots kept him in Newcastle, trying to make him agree to their terms, but Charles could not accept that a king had to negotiate for his throne and spent most of his time playing golf and chess. The Scottish forces had better luck in their talks with Parliament. In early 1647, they handed over Charles and marched back home with a plump ransom.

On the journey from Scotland, Charles and his captors were mobbed by sick citizens, still convinced that his touch could heal the King's Evil. The Puritans mocked Charles's healing touch as nothing but superstition, the king's guards nicknaming him Stroker. But Parliament worried about the hold that Charles had on England's heart. Some wanted to put him back on a diminished throne, forced to accept a stronger Parliament and a Puritan church. Others didn't trust him even that far and wanted him kept in prison indefinitely. Charles did not languish while the two sides tried to come to an agreement; instead, he played the New Model Army and Parliament against each other, seeing which would offer him the best terms.

If these machinations weren't distraction enough, Parliament also decided it was time to crush the rebellion in Ireland for good. The New Model Army, however, saw the war as a ruse for Parliament to destroy its growing independence and mutinied. In its boldest move, a force of five hundred soldiers snatched Charles from Parliament's protection and began negotiating with the king, offering to let him take back his throne if he agreed to the army's terms rather than to Parliament's.

In this chaos, the royalists at Oxford could dig in their heels. Willis's old friends welcomed him back by giving him a medical degree, which before the war would have taken seven more years of study of Galen and Hippocrates to earn. Willis's education during the war had been a few months at most of reading books between fusillades, dabbling in experiments that had nothing to do with a proper medical education, and talking with the followers of William Harvey—perhaps even Harvey himself. His medical degree was a token of gratitude.

Armed with his degree, twenty-five years old, Willis began the hardest chapter of his life. A poor, orphaned soldier on the losing side of a civil war, he would struggle for a decade to survive in the turmoil between Puritans and conspiring royalists, in a country still facing years of war and the execution of a king. For Willis and many like him, these years would feel like a national insanity.

Yet the political chaos also turned Oxford intellectually into a fizzing vial of spirits, a place where alchemists jostled with Aristotelians, where telescopes were trained at the sky and microscopes at the legs of fleas. Scandalously new ideas about the soul itself bubbled up—not theological ideas so much as scientific ones, even political ones. And out of the experience, Willis would find his own ambition to make an anatomy of the soul.

—

The new doctor struggled to find patients. His past was suspicious, his experience thin. He stuttered and handled himself awkwardly.

The diarist Anthony Wood later called him "a plain Man, a Man of No Carriage, little Discourse, Complaisance or Society." Each week Willis traveled to the markets in surrounding towns with his old university companion Ralph Bathurst and a fellow royalist ex-soldier named Richard Lydall. There they competed for business with the quacks and mountebanks. Although Willis had a medical degree, the mark of an elite physician, he still had to work as a "pisse-prophet." People would come to him with the urine of their sick relatives. He would swirl it in a flask, diagnose the malady by its color, and then prescribe a remedy. The few patients he treated—mostly poor royalists who lived in the countryside—usually came to him through some connection from the war, many of them after getting treated badly by quacks, whose remedies Willis called "a sword in a blind man's hand." Few of them could actually pay his fees, but Willis treated them anyway. When he needed to visit patients, he rode a day or more on a horse he shared with Lydall.

Willis had one consolation in his meager practice: it left him plenty of time for scientific work. As an apprentice to Mrs. Iles, he had gotten a taste of Paracelsus and had learned how to prepare medicines in a laboratory. Now he and Bathurst and a few other novice alchemists decided to set up the first laboratories at the university, complete with furnaces, alembics, crucibles, and special ingredients such as mercury, oil of amber, and various "drugges from London." It was dirty, expensive, dangerous work, and Willis would later complain of "the filthiness and soot in which I was involved, being condemned among the metal."

Some of Willis's work was just dabbling. He had a fondness for laboratory tricks, creating invisible ink and figuring out the chemistry that allowed a famous magician of the time to drink water and vomit it back up in a rainbow of colors. But during his time among the metal, he also had some profound realizations. He became more and more disenchanted with Aristotle's four elements, finding them to be empty of meaning. Willis would later write that it would be just as unenlightening "to say a house consists of wood and stone, as a body of four elements." Turning away from Aristotle, Willis delved instead into medical alchemy. Paracelsus had made a deep impression on him as a young apprentice, and now he found inspiration in a new medical alchemist whose work had just emerged after decades of censorship.

Joan Baptista van Helmont was born in 1579, almost forty years after the death of Paracelsus. He attended the University of Louvain (in what is now Belgium), where his Jesuit teachers still relied on Aristotle for their teaching. Just before van Helmont was to receive his master's degree, he asked himself what he had really learned. "It dawned on me that I knew nothing," he later wrote. He abruptly abandoned the university, made Paracelsus his model, and wandered across Europe for years, gathering knowledge. Later he claimed that on his travels he once held the philosopher's stone in his hand. It was the color of powdered saffron, he said, and was heavy and brilliant like broken glass.

When van Helmont returned home to Flanders, he set up a laboratory in which he could come face-to-face with nature. There he worked quietly for fifteen years, emerging from his laboratory only to dispense free medicine to his neighbors. In those years, van Helmont created his own brand of medical alchemy, drawing from Paracelsus, other alchemists, and his own revelations.

He declared that all matter began as water, and that it took on a form when a spiritual seed penetrated it. Hidden spiritual seeds in hot springs created the minerals that built up around them. Life emerged from water in the same way, as van Helmont demonstrated by planting a willow tree. He put a five-pound sapling in a tub of two hundred pounds of earth. For five years he gave the tree nothing but water and then weighed both tree and earth. The tree had grown to weigh 169 pounds, while the earth had lost only a few ounces. "Hence one hundred and sixty-four pounds of wood, bark, and roots have come up from water alone," he announced. The willow was nothing more than transmuted water given form by the willow's archeus.

To liberate an archeus from its crude covering, van Helmont used fire, which he believed revealed the archeus in its pure form as a swirling cloud of smoke. Van Helmont named these mystical clouds "gas." He reasoned that since each archeus was different, fire released a different gas from each one. He succeeded in isolating a number of gases for the first time, including sulfur dioxide and carbon monoxide. Like Paracelsus, van Helmont made scientific breakthroughs not in spite of his mysticism but thanks to it.

To van Helmont, the human body was a hive of souls. Souls governed every organ, sensing the outside world and speaking to one another in the language of sympathy. They carried out their own actions, even experienced pain and emotions. Rigor mortis was the horror that muscles felt at the prospect of death. The body's archei worked like internal alchemists, transmuting matter from one form to another. Van Helmont found that if he fed chickens sharp bits of glass, they passed out of the birds polished. Only acids, the tools of alchemists, could have such an effect. Not just any acid could do the job. Vinegar, for example, was useless. Van Helmont identified the specific kind that dissolved food—hydrochloric acid, which he created by combining sea salt and potter's clay. An archeus transformed food, van Helmont argued, by entering it and exalting it—that is, by fermenting it. The food lost its old identity and could now take on a new one, to become part of its new host. Ferments transformed matter not just in the stomach but in the liver, where the blood of the veins was created, and in the heart, where the spirit of life was created and colored the arteries bright red. Ferments were responsible for all change in nature; they turned dough to bread and lead to gold.

Van Helmont believed that the ferment crucial to life took place in the stomach, separating food into what was good and what should be expelled. The stomach, he reasoned, must also be the home of the primary soul of the body, which issued orders like a monarch, overruling any decisions made by the other souls of the body. The nervous system and the brain paled in comparison. They perceived the world through a few narrow channels, depending on spirits as their messengers.

Not surprisingly, van Helmont did not believe that the soul resided in the ventricles of the head. He did not even think that reason was the noblest gift of humans. Reason was actually a disease of the soul, he declared, a corrupting parasite that distracted the soul from a true union with knowledge. Nor was reason anything special to humans. Wolves could figure out how to corner a dog in its favorite sleeping spot; bees could count. Declaring that man was a rational animal was just the confusion of heathen philosophers.

Van Helmont earned a reputation among the Flemish authorities as a troublemaker. As long as he kept his ideas to himself they left him in peace. But in the 1620s, he drew their ire by stepping into a long-running debate over how to cure wounds from weapons. A Paracelsist philosopher explained that a salve made of oils, fat, and other ingredients should be applied to the blood left on the weapon. Thanks to its long-distance sympathy with the wound, the salve could heal it, no matter where the victim might be. A Jesuit professor condemned the idea as "the devil's deceit," but van Helmont thought a weapon-salve made perfect sense. Everything in his world contained a soul that could send out a magnetic-like sympathy to everything else. The salve could waken the blood's soul and let it perceive the wounded victim far away. It was not magic or the work of the Devil, van Helmont declared, but instead the sympathy of like to like, of nature itself.

The Spanish Inquisition declared him suspect of heresy, and two years later, his tract on the weapon-salve was impounded. After years of interrogations and court proceedings, he was convicted of "perverting nature by ascribing to it all magic and diabolical art," and was placed, like Galileo, under house arrest. He stayed there for the rest of his life, forbidden to publish a word.

Van Helmont saw his persecution as the work of God, who "suffered this Evil and unprofitable Servant to be sifted by Satan," but he also believed that God had chosen him to discover the true workings of nature. He had a duty to write down what he had discovered. "I knew perfectly that the hand of the Lord had touched me," he wrote, "and therefore, in a full tempest of persecutions, I wrote a Volume." Not long after he finished it, on the evening of December 30, 1644, van Helmont asked for the last rites of the Roman Catholic Church.

Shortly before his death, van Helmont asked his son Franciscus Mercurius to ensure that his book was published. It was printed in Latin in 1648 and in English two years later. Puritans admired the scorn van Helmont poured on Aristotle, the faith he had in revelations from God and in hard laboratory work. For Puritan alchemists, van Helmont was the new model of both science and holiness. One English admirer declared that he had been sent to Earth by God, his system "ordained in these last times by especial providence of God, for the comfort and relief of distressed Man." Willis was no Puritan, but he recognized van Helmont's greatness and abandoned Galen's claim that food was cooked in the stomach in favor of van Helmont's acids and ferments.

Galen's version of the body was now under attack from all directions. Anatomists exposed his errors as well. The French anatomist Jean Pecquet discovered that when food was digested in the stomach and intestines, the resulting chyle traveled directly into the surrounding veins and bypassed the liver altogether. The Cambridge professor Francis Glisson examined the blood vessels of the liver closely and could see that they were designed to discharge bile, which was picked up by other vessels that carried it to the gallbladder or the intestines. The liver, in other words, was not the home of a soul or the author of blood. It was simply a filter.

In a description of Pecquet's work, an anatomist wrote an epitaph for the liver:

Pause traveler!
There is buried in this grave he who buried many;
The prince of your body, its judge and its cook:
The Liver.

The vegetative soul was gone, and now it seemed as if everything was open to question. If blood was not perfected in the liver, then where? What made the body warm? Where did it find the energy to move itself?

Willis looked for answers in his own laboratory. He boiled the blood and urine of his patients and added solvents to break them down to their components. He named the substances he isolated water, earth, salt, sulfur, and spirit—borrowing Paracelsus's labels. If blood could be broken down into these components, Willis wondered, did that mean that humors—the bedrock of medicine—were not actually things in themselves?

Willis looked beyond the body to find the substances that might be essential to it. Alchemists had known for decades that there were mysterious links between gunpowder and life. Gunpowder's essential ingredient, saltpeter, could be extracted from manure or the boiled-down remains of crayfish. Some alchemists suspected that saltpeter even existed in the air, where it clashed with sulfur rising from the Earth to create thunder and lightning. Several philosophers had suggested that our bodies might move thanks to an inner violence as well. Aside from gunpowder, alchemists knew of only one other substance that could explode, an olive-green powder known today as gold hydrazide. In Willis's time it was called *aurans fulminans,* or exploding gold. Alchemists had experimented with it for centuries, and Willis created some for himself. It needed no fire to explode. He would place a little powder on a spoon, cover it with a heavy coin, and gently tap it on the tabletop. The *aurans fulminans* would blow the coin up to the ceiling with "a vehement crash." Perhaps, Willis speculated, the body was warmed by this sort of flameless explosion.

—

In 1647, Parliament finally began to sweep Oxford clean. After months of negotiations, Charles had secretly opened talks with Scotland, and Parliament worried that royalist bastions like Oxford might rise up. An official Visitation came to the university to test its loyalty. One by one they summoned dangerous students, fellows, and deans and asked each one the same question: "Will you submit to the authority of Parliament in this Visitation?"

Samuel Fell greeted the Visitors with a three-hour sermon, denouncing them and claiming allegiance to the king alone. The

Visitors had him arrested. Other royalists were hauled out of their rooms. The less prominent tried to put off the Visitors without so much public defiance. One student named Robert Whitehall responded to the grilling by declaring, "My name is Whitehall, God bless the poet, / If I submit the king shall know it." Those who did not submit were thrown out of the university, stripped of their privileges and positions. All told, about half the university, somewhere between three and four hundred members, was expelled.

Thomas Willis remained. He was too insignificant to raise the suspicion of the Visitors, and so they did not summon him to take an oath. The threat of the oath hung over his head for years, but Willis could not leave Oxford. He had nowhere else to go, his beliefs prevented him from swearing falsely, and so he became invisible. Willis withdrew from the official life of the university, keeping busy instead with his medicine and his alchemy. He would secretly defy the Puritans, and in time become a hero to those who still swore allegiance to the king.

The men who were brought in after the purge disgusted Oxford's royalists. Samuel Fell's son John later wrote, "An almost general riddance was made of the loyal University of Oxford; in whose room succeeded an illiterate rabble, swept up from the plough tail, from shops and grammar schools, and the dregs of the neighbor University." Anthony Wood dismissed most of the newcomers as "harpies and stinking fellows." But there were glaring exceptions among them—most exceptional of all being John Wilkins, the broad-shouldered new warden of Wadham College, with a reputation for "a very mechanical head."

Wilkins came from a solidly Puritan family and had studied at Oxford under Laud in the 1630s. As a student, he had delved into mathematics and astronomy, getting an education in both Copernicus and Galileo, and had written a series of best-sellers in plain but graceful English, introducing his countrymen to their new solar system. The moon, he revealed, was not some astrological fixture in the firmament, but a craggy ball of stone. Wilkins hoped people might some day journey to it in a flying-chariot "in which a man may sit and give such a motion unto it, as shall convey him through the air." The Earth, far from being the center of the universe, was a planet just like Venus or Mars, revolving with them around the sun, which in turn was one star among many in the universe. Wilkins did not convince everyone—one critic in 1646 wrote a pamphlet called *The New Planet No Planet, or The Earth No Wandering Star Except in the Wandring Heads of Galileans.* Wilkins never sank into bitterness; he was always amiable and gracious. "If there be nothing able to convince and satisfy the indifferent reader," he wrote in the dedication to one of his books, "he may still enjoy his own opinion."

Wilkins left Oxford to serve as minister to a series of aristocrats, some of them moderate Puritans, others moderate royalists. For a time, he worked for the German prince Charles Louis, a nephew of King Charles I living in exile in London. By accompanying the prince to the Continent, traveling through the devastation left by the Thirty Years War, he could see for himself what fanatic intolerance could wreak. "It is an excellent rule to be observed in all disputes," Wilkins wrote, "that men should give soft words and hard arguments; that they would not so much strive to vex, as to convince an enemy."

While working in London for Charles Louis in the early 1640s, Wilkins joined forces with a handful of other natural philosophers to create a scientific culture. Nothing like it yet existed in England. In the late 1500s English noblemen who developed an appetite for curiosities and marvels, filling their cabinets with strange stones and mechanical birds, came to be known as virtuosi. The statesman Francis Bacon struggled for years to persuade Queen Elizabeth to turn this dilettantism into national institutions for promoting knowledge—a zoo, a botanical garden, a vast library, a chemical laboratory. His lobbying came to nothing, and James I proved equally indifferent. Bacon turned to writing manifestoes instead, in which he dreamed of a new science, of natural philosophers gathering evidence without the blinders of obsolete dogma, communicating with one another about their discoveries, testing hypotheses, and creating a utopia. Bacon's words went pretty much ignored until after the English Civil War, when men like Wilkins grew proud to call themselves virtuosi and tried to make Bacon's dreams real.

Wilkins fostered good connections with Parliament, and was rewarded with an Oxford post. Immediately he set out to lure great thinkers to Oxford and became friends with local virtuosi such as Willis. Wilkins set no religious or political standard for membership in his circle, which he drew from both sides of the Civil War. It included the astronomer Seth Ward, a dedicated royalist, as well as John Wallis, a mathematician who had served as Parliament's code-breaker. Wilkins shepherded bright students into the university from even the staunchest royalist families, including a sixteen-year-old prodigy named Christopher Wren.

Wren came to Oxford freshly shattered by the war. Under Charles I, his father had served as dean of Windsor, a powerful church position that came with wealth, lands, and a grand house on the grounds of Windsor Castle. Dean Wren's loyalty never wavered in the months before the war; he followed his king to the north of England and then settled in Oxford. For his loyalty, Parliament stripped him of his lands, and when their soldiers stormed into Windsor Castle they took all of the Wren family's possessions except a few books and a harpsichord.

During these calamitous years, young Christopher Wren found a refuge of sorts as a student at Westminster School in London. But after the fall of the king's army, his father took him along with the rest of his family to a small village near Oxford. There they lived in obscurity with Christopher's older sister and her husband. The ex-dean filled his time tutoring his son, imparting to him a passion for mathematics and astronomy. In some ways it was an outmoded passion—in the margins of a book on astronomy, Wren's father scribbled, "Either God or Copernicus speaking Contradictions, cannot both speak truth."

Even in these humiliating straits, Christopher Wren quickly proved himself a prodigy. He mastered spherical trigonometry at age fifteen, created sundials, and built beautiful pasteboard models of muscles. One of his mentors declared, "I can shortly expect great things, and not in vain." Wilkins, an old friend of the family, recognized Wren's remarkable intellect and promised to protect the frail, politically suspect boy at Oxford. At the university, Wren began working for Wilkins, who became a second father to him, and a more up-to-date one at that.

Wilkins, Wren, Willis, and their fellow virtuosi came together to form what they called the Oxford Experimental Philosophy Club, which met every Thursday afternoon to watch experiments and debate their meaning. Everything in science was within their reach, if not their grasp. They tried to invent a universal language free of all confusion. They grew plants from the New World and the tropics in a "physik garden" to see what uses they could find for them. They created counterfeit rainbows and talking sculptures, designed submarines and spacecraft. Wilkins persuaded Wren that Copernicus was right and inspired the boy to prove some of Wilkins's own speculations true. (Wren called himself "a most addicted client" of Wilkins.) Wilkins had written that the moon had a landscape of its own, and Wren began to design new lenses and telescopes to map it. He used the same methods to make new microscopes, through which he looked at a minuscule world that only a handful of people on Earth had yet seen, gazing at the hairy legs of fleas and the stained-glass leading of a fly's wing. He made a compass needle as long as a sword to measure the Earth's quivering magnetic field. Wren even showed promise as an architect, building intricate beehives out of glass, which he ornamented with sundials, statues, and weather vanes.

Wren, Wilkins, and the rest of the Oxford circle also helped Willis bloom as a natural philosopher. The club referred to him as "our chymist." Willis learned a great deal from the astronomers and mathematicians, but he was influenced most of all by an anatomist with goose-gray eyes named William Petty, who joined the circle in 1648. Although Petty was only twenty-five, he had far more to teach Oxford's professors than they him. One of them declared, "He was born with three-fourths of what he knows."

Petty was not typical Oxford material. His father was a failed clothier. As a boy, he was fascinated by the mechanical arts, gawking at watchmakers, blacksmiths, and carpenters. At age thirteen, he left home to serve as a cabin boy on a merchant ship shuttling across the English Channel. He learned how to read a compass and track tides, but Petty's greatest gift was turning nothing into something, buying fake jewelry in France and selling it to lovesick men in England. The money he earned saved him when he broke his leg and the ship's crew unceremoniously dumped him on the coast of Normandy. Although only fourteen, Petty was able to pay an old woman to set his leg and an apothecary to sell him crutches. He then approached the Jesuits who ran a school at Caen and persuaded them to admit him, offering a Latin poem he had written for the occasion to prove his talent. To pay his way, he traded in beeswax and tobacco pipes.

Petty returned to England to serve in the navy before making his way back to the Continent, where he studied medicine in Leiden. At the time, Leiden boasted one of the most advanced medical schools in Europe and one of the few places where Harvey's work was taught. There Petty learned not only about the circulatory system but also how to dissect a body in order to investigate its workings. After a year in Leiden, he made his way to France, arriving in Paris in 1644, hoping to study at its anatomy school. Once again Petty had no money—for a week he survived on a few pennyworth of walnuts—but he managed to meet some of Paris's greatest thinkers, including an aging English tutor with a bald head and a palsied hand, Thomas Hobbes. Known to only a few Englishmen at the time, in less than a decade Hobbes's name would become a universal insult.

For most of his career, Hobbes worked as tutor to the family of the earl of Cavendish. It was only in his forties that he got around to reading Euclid for the first time, and he fell in love with the unquestionable logic of geometry. He thought all philosophy should find the same certainty, even the philosophy of human nature. His work took Hobbes to Europe in the 1630s, where he became friends with Mersenne, Gassendi, and Galileo, whose ideas led Hobbes to conclude that the universe was purely mechanical, comprehensible only by precise measurement of its motions. Hobbes saw no limit to this method. "For what is the heart but a spring," he wrote, "and the nerves but so many strings; and the joints but so many wheels, giving motion to the whole body, such as was intended by the artificer?"

Hobbes plunged human nature completely into this materialist tub. "It is manifest," he wrote, "that the immediate cause of sense or perception consists in this, that the first organ of sense is touched and pressed." Seing a sunset meant that vibrations traveled from the sun through the air, pressing against the eye. Particles within the eye were put in motion, which in turn set up a motion in the "brain or spirits or some internal substance of the head." Hobbes didn't much care what exactly moved inside the head as long as it moved. The brain pushed back in the direction from which the outside motions came, creating the impression of the sun where it actually was in the sky.

But to Hobbes, the brain was only a way station in the path of perception. Motion propagated down from the head to the heart, which Hobbes called "the fountain of all senses." While he was opposed to much of what Aristotle had to say, he accepted this key piece of philosophy. "If the motion be intercepted between the brain and the heart by the defect of the organ by which the action is propagated," he declared, "there will be no perception of the object."

Just as perceptions were nothing but matter in motion, so too were memories, learning, and passion. Hobbes claimed that when the brain was put into a certain motion, it put the heart in motion as well. If the motion impeded the flow of the blood around the heart, it caused pain; an easier flow brought pleasure. Motion could communicate in the opposite direction as well, from body to brain. Stirring up the spleen caused fearful dreams, while the heat of the heart generated anger and images of enemies.

Hobbes launched his philosophical project years before

Descartes published his own work, and yet both men wound up with much the same picture of mechanical life. As soon as Hobbes had an opportunity to read Descartes's *Discourse,* however, he knew there was a crucial difference between the two of them. Descartes took the soul out of the material universe but left human beings with an immaterial soul capable of reason and of intervening between sensation and action through the pineal gland. Hobbes on the other hand did away with the immaterial altogether. The notion of immaterial spirits was nothing more than a delusion of human imagination. Reason was not the work of an immaterial soul; it was simply the body's ability to keep names for things organized, to keep its thoughts in an orderly dance.

Just as atoms moved according to certain laws, Hobbes believed people were driven by two overwhelming passions—their appetites and their aversions—on which all human emotions were variations. All human beings moved away from what is evil for them and toward what is good for them with an inevitable predictability, "no less than that whereby a stone moves downward," Hobbes wrote. "I put for a general inclination of all mankind a perpetual and restless desire of power after power, that ceaseth only in death." It is our fear of death, a property of how we are constructed, that makes us rational.

From atoms to humans, so from humans to nations. With a science of the soul in hand, Hobbes was confident that he could create a science of the state. In both cases, he needed only to understand how the individual parts worked. When people live without a state, they are prey to their passions. They all compete with one another and wind up with lives that are, in Hobbes's most famous words, "nasty, brutish, and short." Thanks to human nature, life outside the state is guaranteed to produce the greatest misery, but by using reason, humans can compose laws that create happiness instead. There is nothing divine about laws, according to Hobbes, because they are as arbitrary and man-made as words. They can be effective only if people agree to be ruled by a sovereign, one who rules over them with absolute authority.

As radical as Hobbes's philosophy was, he didn't think of himself as a revolutionary. Hobbes traveled in royalist circles and once even stood for Parliament as a royalist. He looked at Puritans with a conservative's disdain, considering their religion nothing more than a folly that made its victim think that the laws of God were revealed personally to him. And it was as a royalist that he panicked in 1640. A short treatise of his ideas was circulating among his friends, just as Charles's close advisors were being arrested and executed. Fearing that his writing would be seen as an attack on Parliament, Hobbes fled to Paris.

In exile, Hobbes took his place again in Mersenne's cell, where he attacked Descartes with remarkable self-importance. Descartes responded with typical contempt: "For unless I am very much mistaken," he wrote to Mersenne, "he is aiming to make his reputation at my expense, and by devious means."

The other English exiles who fled to France looked on Hobbes with a mixture of admiration and suspicion. As far as they could tell, his political tracts seemed to support the monarchy over the rabble, but he could also come across as an atheist. When Prince Charles fled to Paris in 1646, Hobbes was appointed his tutor, but he was permitted to teach his student mathematics and nothing more.

During his exile, Hobbes took the young William Petty under his wing. They talked for hours about philosophy in Hobbes's library (which, Petty noticed, had only half a dozen books). Petty drew illustrations of light rays for a book about vision Hobbes was writing. Hobbes was also about halfway done with a book that would explain the body's machinery. Petty helped him through Vesalius's *Fabrica,* which he hoped would show that nerves were arranged around the heart—anatomical evidence that pain and pleasure were caused by the movement of the vital spirits surrounding it. Vesalius did not give him the answer he wanted, so Hobbes attended Petty's dissections, peering into the open chests of the dead, hoping to see the strings of happiness and misery.

Hobbes left a deep impression on Petty, who later wrote that he was the one writer "who always examines everything he talks about with the most complete care." In 1646, when Petty left Paris, Hobbes gave him a microscope, but Petty came away with much more. Hobbes had introduced him to the mechanical philosophy and to a way of looking at man that would guide his work for the rest of his life. Petty now saw the body as an intricate machine that could be studied, like a watch, as an assembly of moving parts. People themselves were like the parts of a watch, he concluded, combining together into nations. Just as anatomy could help a doctor preserve health, the study of people could be used to turn a profit. Six years after he left Hobbes's company, Petty would put those principles to practice on an entire country.

When Petty returned to England, he was still penniless, in spite of his peerless education. In London, as in Paris, he worked his way into the leading intellectual circles, but now, instead of Catholic priests and royalist exiles, he hobnobbed with Puritan dreamers and Paracelsist utopians. Petty offered his own proposal for a teaching hospital to be set up on Francis Bacon's principles, for the study of things rather than the rabble of words. Meanwhile Petty struggled to find a way to turn his intellect into money. He was full of inventions—a mechanical corn planter, a bridge without supports, a double-writing machine—but nothing came of any of them.

Petty's medical training finally brought him a living. In 1649 he traveled to Oxford to get a medical license, bringing with him the microscope Hobbes had given him, his dissection tools, and his revolutionary ideas. Moving into rooms above an apothecary on the High Street, he had easy access to the ingredients he needed for running chemical experiments and making remedies. He got his degree by presenting letters from the commander of the local garrison and soon was cutting open cadavers as assistant to the professor of medicine, Thomas Clayton. After a few months, Clayton left the anatomy work to Petty completely. Politics for once probably had little to do with the promotion. Clayton was simply a terrible anatomist. "Being possessed with a timorous and effeminate humour," the diarist Anthony Wood wrote, Clayton "could never endure the sight of a mangled or bloody body."

Petty had no such qualms. For the first time in Oxford's history, an anatomist skipped Galen entirely and cut open the cadavers of criminals himself. (Oxford anatomists had the right to dissect anyone executed within twenty-one miles of the university.) Petty anatomized animals as well, not only for his lectures but in his private rooms. He much preferred the company of Willis and Wilkins and their friends to dry academic rituals. (He complained that universities "sought truth as courts do justice.") Petty took over as host for the Oxford circle's weekly meetings. In one experiment, a fat boy sat on a stool below which the bladder of an ox was placed. A man blowing into the bladder through a pipe was able to lift the boy two inches off the ground. How the air could be so strong no one could say.

In Petty, these natural philosophers met a kindred spirit. He would begin his dissections with a poem in praise of Harvey, who in 1649 still had few supporters in England. But Petty was also deeply influenced by the mechanical trio of Hobbes, Gassendi, and Descartes. Life for Petty was now a machine to be dismantled. "The most mysterious and complicated enginry is nothing to the compounded and decompounded mysteries in the fabric of man," he wrote. "All their static and hydrostatic, their hydraulic and trochaulic, thermoptic and scenoptic, their recouistics and music, their pneumatics and ballistics and all their other mechanics whatsoever, are no more compared to the fabric of an animal than putting two sticks across is to a loom, a clock, or a ship under sail—the latter whereof, supposing her men to be animal spirits, comes nearest to an animal of anything I know."

Petty and his new associates at Oxford had some reservations about the mechanical philosophers of Paris, though. Descartes was "a most excellent mathematician," Petty wrote, but he complained that Descartes based his ambitious claims on few experiments: "I cannot believe his principles of Philosophy to be so firm that stand upon such narrow feet as those few experiments mentioned in his works." As a result, he was not terribly eager to learn from Descartes, writing that he "never knew any man who had once tasted the sweetness of experimental knowledge, that ever afterward fasted after ye Vapour garlick & onions of phantasmatical seeming philosophy."

—

The Oxford circle worked together even as their country threatened to pull itself apart. In 1648, a great force of royalist soldiers and Scottish allies came together in the north of England to fight for the king. Cromwell marched north and routed them, driving the Scots into a muddy retreat. A few hundred royalists held out in the city of Colchester for a while, but Fairfax besieged them until they were reduced to killing cats for food.

Parliament negotiated with Charles on the Isle of Wight for a settlement, but tempers were running short. The king was now looking more like a tyrant who slaughtered his own subjects and had to be dethroned. Some generals even talked of putting him on trial for treason. In a moment of exasperation, Cromwell declared, "I tell you we will cut off his head with the crown on it."

Ultimately, they did. In January 1649, Charles was brought to Westminster Hall and accused of treason, tyranny, and murder. He refused to recognize the court's authority. "I would know by what power I am called hither," Charles asked. His accusers claimed the power of God. Charles was not impressed, and for an entire week he refused to respond to their questions. Eventually Cromwell and the other judges went ahead on their own and sentenced him to death. On a bitterly cold day, Charles was led to a scaffold at Whitehall, where once he had held lavish masques celebrating his divine right to rule. Soldiers blocked him from the view of onlookers, who had climbed onto nearby roofs. Charles's only audience was the few scribes appointed to record the execution. "I am the Martyr of the people," he told them. He laid his head on the block. It was he, and not his judges, who gave the signal to the executioner to bring down the axe.

Great Britain was now a republic, with Cromwell its first chairman of the Council of State. But even with the king dead, the fighting was not over. Over the next four years, Cromwell led Parliament's armies against radicals and mutineers and sailed to Ireland to crush a rebellion there. After Charles I's execution, his son Charles assumed his crown in exile and allied himself with the Scots, who were enraged that Parliament had killed a Scottish king. In 1650, Charles marched south into England; Cromwell promptly stopped the invasion, sending him scurrying back to Europe.

Eight years after Charles I had raised his standard against his own Parliament, Parliament had executed him, humiliated his son, crushed revolts in Scotland and Ireland, and turned Britain into a republic. For the royalists, it was a defeat far more devastating than they had ever imagined. Their old king was dead, their new king sent running, their cathedrals were empty and vandalized, their bishops executed or under virtual house arrest. To them England's madness seemed now as if it might never end.

—

In the years after the death of the king, Thomas Willis was intimately aware of this madness. As he rode out of Oxford, he passed the burned husks of houses gutted by the Civil War's fires, the crumbling earthen walls, and the new slums rising up in their place. He traveled out into the countryside, where he visited patients—mostly poor, disempowered royalists—and at the end of each day he would write down notes about his cases in precise, minute Latin, trying to make some sense of their complaints.

One of his casebooks has survived, a small brown leather volume that Willis must have slipped into a pocket each morning. It offers a rare glimpse into his practice in the early 1650s, describing afflictions that can be recognized today as tuberculosis, appendicitis, intestinal worms, kidney stones, malaria, and arthritis. Sometimes the cases are harder to pin down, such as that of a man Willis once encountered who "was wont every day, two or three times, for about two hours, continually to belch, with such a noise that he might be heard far and near, at a great distance."

When Willis tried to diagnose these patients in the early 1650s, he still tended to think about diseases as imbalances or corruptions of the four humors. When he treated an infant sick with measles, he observed that the child's mother "was night and day almost continuously morose and afflicted herself beyond reason by not sleeping or eating. It was also inevitable that the milk taken by the infant was infected with a melancholy juice or black bile." His treatments were often straight out of Galen, with plenty of blood-letting and purging. In September 1650, for example, Willis examined a fifty-year-old countryman who had been spitting up blood for four months and had begun to cough up putrid phlegm. Unaware that the man had tuberculosis and that bacteria were destroying his lungs, Willis diagnosed him instead as having "sharp blood" that was rising into the lungs and eroding them. He had the man bled from the arm with leeches and prescribed a roasted apple that had been hollowed out, stuffed with frankincense and sugar candy, and cooked in ashes, along with the milk of an ass mixed with rose-water and a bowl of turtle soup.

Some symptoms Willis could not account for with Galen's medicine. A forty-eight-year-old noblewoman "of melancholic temperament" had suffered for years from grief, but had recently developed more severe symptoms. Heat and sweat would spread over her head, and once these passed, she broke out in scales and oozing pustules. She felt pains creep within her body, down from her head to her back and shoulders and down into her fingers. Sometimes she felt strange movements inside her abdomen, "such as that of a wild animal running from the side to the ilium, but then she felt it no more." Although she had a healthy appetite, she couldn't tolerate the slightest draft of air; she kept her house closed up and a fire burning in summer.

"It will not be an easy manner to name the proper causes of this state, which is not to be sought in the treasury of ancient medicine," Willis wrote in his casebook. He decided he would have to think about the woman chemically. He speculated that vapors might be rising from her humors because some sulfurous or nitrous substance had been added to them; perhaps the humors were coagulating or dissolving or otherwise changing in the same way Willis's chemicals changed in his laboratory. "If we plough with this calf," he wrote, "the enigma of these prodigious conditions may perhaps, in some way, be solved."

Neurological and psychological disorders were even greater enigmas for Willis. As a young doctor, he struggled to understand how patients might become terrified of cats or toads or some particular kind of food. In spite of his confusion, he observed this sort of condition with exquisite care. In his notes and in the books that later grew out of them, we come across the first known clinical descriptions of many neurological and psychological disorders. (Of narcoleptics, for example, he wrote, "Whilst talking, or walking, or eating, yea their mouths being full of meat, they shall nod, and unless rouzed up by others, fall fast asleep.") In June 1650, Willis wrote that "a countrywoman, aged about 45, for long melancholic, was seized by mania on the 29th of June, so much so that it was necessary to bind her with chains and ropes to keep her in bed. On the fifth day 1/2 pint of blood was drawn from the basilic vein. At bedtime she took 2 grains of laudanum in a coction of barley with an infusion of poppy flowers dissolved in a sweetened cardiac syrup. She slept about 3 hours that night. Early in the morning she had 3 stools from the enema administered the day before. About noon she slept again. In the evening I visited her. She was now shouting wildly, now singing, now weeping." The next day she died. Willis had written the earliest known clinical case-study of manic-depressive psychosis.

Willis sometimes encountered patients who had lost control of their body in any number of ways. "I have known some, who had all the muscles and tendons through their whole body afflicted with contractions and leapings without intermission," he later wrote. "I have known others whose thighs, arms, and other members, were perpetually forced into various bendings and distortions. And also others I have seen, who of necessity were compelled to leap and run up and down, and to beat the ground with their feet, and hands, and if they did it not, they fell into cruel convulsions."

Somehow, people could lose control of their bodies, even when they were perfectly sensible. Once, for example, Willis was asked to treat a tormented sixteen-year-old girl. "At first," he wrote, "she was troubled, though not in any grievous manner, with a headache, and giddiness for many days; then she felt now in one of her arms, and then in another, a trembling and sudden contraction; which kind of convulsions, returning often that day, endured scarce a moment; the next day sitting nigh her sister in a chair, suddenly leaping out, she fetched one or two jumps, and many others successively; with wonderful agility, at the distance of many feet; then, when she was come to the farther part of the chamber, she stood leaping a great while in the same place, and every time to a great height; when her legs were quite tired with leaping, she fell on the floor, and presently she flung her head here and there with wonderful violence, as if she should shake it from her neck."

When the girl became too tired to carry on this way, she would punch and kick the walls and the floor. "When by reason of shame or modesty, due to her friends and bystanders, she did hinder herself with great violence from these motions (for all the while she was herself, and spoke soberly) the distemper being sent inwardly, she was very much infested with a mighty oppression of the heart, with a bemoaning and very noiseful sobbing; then when she would be as herself, she was forced presently, the fury being transmitted to the muscles of the outward limbs, either to leap about, or to fling here and there cruelly her head, or arms, or also to run about the chamber, most swiftly, or to beat the ground with her feet, because these kind of vehement motions of her limbs or viscera, in the tragedy of the distemper, did mutually relieve themselves, returning as were in a round."

According to Willis, "some thought possessed her with an evil spirit." Many of his contemporaries saw convulsions as the work of demons, but Willis did not prescribe an exorcism. Instead, he instructed that she should be given an emetic to make her vomit, for ten ounces of blood to be drained from her arm, and for her to swallow "antidotes of the powders of precious stones, of human skull, and the root of peony."

Willis sometimes tried to understand these disorders by falling back on Galen. He wrote in his 1650 casebook that the symptoms of hysteria "owe their origin to the uterus." As humors built up in the womb, they created foul vapors that rose up and struck many parts of a woman's body. But Willis also made careful note of how chemistry—in the form of things people ate and drank—could directly affect their behavior. One night, he stopped at a country inn to spend the night. He was asked by the owner to examine two peasants, father and son, who were "overwhelmed with the most profound sleep." Willis went, "not only out of charity, but led also by curiosity." In their cottage the two men were slumbering, as they had been since the day before. It turned out they had accidentally eaten poisonous henbane, having mistaken it for parsnips. Willis thrust a feather down their throats to wake them up. The following day both father and son returned to themselves.

Eating and drinking could not only send a person to sleep, but drive sleep away, as Willis observed when the first coffeehouse opened in Oxford in 1650. The house attracted Willis's circle of friends, who went to talk natural philosophy or to get gossip from London, and as they drank, Willis observed how the coffee altered them. "The drinking of coffee (in use formerly among the Arabians and Turks)," he wrote, "which is drunk by our countrymen, either physically or out of wantonness, all sleepiness being driven away, doth produce unwonted waking and an unwearied exercise of the animal faculty; that some have a necessity to study late in the night, or presently after drinking, or a full meal, by drinking a due quantity of this liquor become still waking, and perform any hard task of the mind, without sleepiness."

Even the loftiest faculties of the human soul could become diseased. His patients, for example, could become completely deluded about themselves: "Some have believed themselves to be dogs or wolves," he wrote, "and have imitated their ways and kind by barking or howling; others have thought themselves dead, desiring presently to be buried, others imagining that their bodies were made of glass, were afraid to be touched lest they be broke to pieces."

As for human memory, Willis learned how vulnerable it could be when he walked into William Petty's rooms one day in 1650. A coffin lay on the floor, inside which was the body of an executed woman named Anne Greene. "She was fat and fleshy, of an high sanguine complexion," Petty later wrote. In life Greene had been servant to Sir Thomas Reade, a local gentleman. Reade's grandson had seduced her, and four months later, as she turned malt in her master's barn, she was overwhelmed with pain. She ran into Reade's office house, and there, to her surprise, she delivered a baby. She then discovered that the infant was dead. Horrified and frightened, she hid the little body in the attic of the house. It was soon discovered, and she was arrested and tried for murder. Her judges sentenced her to hang.

Greene languished in prison for three weeks until the cold, rainy day of her execution. That morning, she left her cell and walked into the courtyard of Oxford Castle. She climbed a ladder, and a psalm was sung. She proclaimed her innocence to the crowd, railing against the lewdness of her master's house. Her executioner put the noose around her neck and turned her off the ladder. She fell and then hanged for almost half an hour. "And while she was hanging," Petty wrote, "divers friends of hers and standers by, some hung with their whole weight upon her, others gave her great strokes on the breasts; and moreover a soldier did the same several times with the butt end of his musket."

Greene's body was cut down and put into a coffin that had been sent by Petty and Willis. When it arrived at Petty's house, relatives and friends of Greene, along with assorted onlookers, were crowded into the dissecting room. Petty had helped to make autopsies a familiar event in Oxford, and they were treated as social occasions. But when the coffin was opened, Petty wrote, "she rattled in the throat; Whereupon a lusty fellow standing by stamped upon her breast and stomach several times with his foot."

The two doctors arrived in the room at that moment and pushed the lusty fellow aside. They tried to revive Greene as she lay in her coffin. "We wrenched open her teeth which were fast set, and put in some strong waters," Petty wrote, "whereupon she seemeth obscurely to cough or spit." Willis and Petty could feel a faint pulse. They rubbed her hands and feet for fifteen minutes, bled her, rubbed turpentine over the rope burn on her neck, and tickled her throat with a feather. They put Greene in a bed and had a woman climb in alongside her to rub her gently. Greene opened her eyes. "We called to her to know whether she could either speak or hear," Petty wrote, "but she could neither." They continued to bleed her and to give her potions made from ground-up mummies and rhubarb and spermaceti. The audience by now had grown into a choking crowd, and so Petty and Willis moved Greene into a small room, where she could sleep through the night. The following morning, she asked for beer to drink, and five days later she was out of bed, eating chicken wings.

The justices wanted to hang Anne Greene again and do it right this time. Petty and Willis pleaded to them to spare her life. They declared that the baby had been dead when it was born and was so small—"a lump of flesh"—that it couldn't possibly have survived even if it had been born alive. Her revival, they said, was a sign of divine providence pointing to her innocence. Far from another execution, they demanded that she should get some compensation from young Reade, who had brought her so much misery.

The doctors won Greene her life and even arranged for her to make some money from the experience. People paid a fee to file into Petty's chambers and look at her lying in her coffin in the very room where she had been about to be dissected. She eventually went home, one newspaper reported, "with the coffin wherein she lay, as a Trophy of her wonderful preservation."

The news of Anne Greene's death and resurrection spread far, and Petty squeezed all the glory he could from it. He wrote to a friend that "my endeavours in this business have bettered my reputation." He even had pamphlets printed in London describing the miracle he and Willis had accomplished, complete with thirty-seven poems written in their honor by their Oxford friends, including Christopher Wren. (Most sang of Orpheus or used other lofty images, but one poet wrote, "Thus 'tis more easy to recall the Dead / Than to restore a once-lost Maidenhead.")

Anne Greene brought Willis some fame as well, but his practice didn't improve much. He would still be casting waters as a pisse-prophet for several more years. The most valuable part of the experience for him was the glimpse he got at the nature of the human mind. When Greene came to herself in bed, she suddenly began to speak, delivering her speech on the gallows again "as if there had been merely a cessation of her life," Petty wrote, "and that she had just gone on where she left off before."

Petty and Willis asked her what she could remember of what had happened to her. She offered no grand visions of the afterlife, no hellish howlings or celestial melodies. She remembered taking off her bodice—which had cost her five shillings, she pointed out—and handing it to her mother. But beyond that, Petty wrote, "she remembered nothing at all; neither how she came out of the prison, walked up the ladder &c, nor what she had spoken on the scaffold; although she had spoken sensibly enough there." According to the pamphlet account, "her memory was like a clock whose weights had been taken off a while and afterward hung on again." Memory was part of the mind's machinery, a part that could falter and fail like all the others.

—

For four years Petty and Willis worked side by side. In that time, Petty helped transform Willis from a shabbily trained physician with a talent for alchemy into a new scientist. He instructed Willis in the blood-and-guts details of dissection. Under Petty's influence, Willis began dismantling his patients after they died, looking for the causes of diseases within. In one dead patient, Willis saw that the lungs were floating in a pool of liquid; he brought the fluid to his laboratory and cooked it, observing that "there was, in the bottom of the vessel, a residuum like mucus or flour porridge and whites of egg, viscous and smelling abominably." Petty also familiarized Willis with the mechanical philosophies of Descartes, Hobbes, and Gassendi. Willis himself would later write that anatomists should learn how the body's parts work, "as in mechanical things when any one would observe the motions of a clock or Engine, he takes the machine itself to pieces to consider the singular artifice, and doth not doubt that he will learn the causes and properties of the Phenomenon."

As closely as they worked during those years, Willis and Petty remained two very different men—Willis the provincial farmer turned soldier turned alchemist turned local doctor, Petty the cosmopolitan philosophical businessman. After four years at Oxford, Petty left to join the English army in Ireland as its physician-general. He organized the army hospitals to fight dysentery and plague, but he also found time to get involved in the real business of the army: turning over two million acres of seized Irish property—two-thirds of the country—to thirty-three thousand English soldiers and investors. Handing over the land to English claimants had become a nightmare, with little clear evidence of who owned what after twelve years of war and famine. A surveyor-general was struggling to find out, but the results were muddled and slow. In the meantime, English soldiers were given scraps of paper in lieu of land, which some of them sold to speculators in order to survive. Cromwell did not want thousands of angry landless soldiers threatening his plans for peace back in England. He needed them settled as quickly and peacefully as possible.

Petty offered to make his own survey of the country, promising to set aside forty acres of good land for every soldier. He would go beyond the original plan for the survey and map Ireland's towns and parishes and baronies, its rivers and ridges and bogs. And he would finish it in just over a year. In devising the survey, Petty proved to be a fine student of his old teacher Thomas Hobbes. Having been taught that the human body was a device made of parts and particles, Petty now saw Ireland as a machine with people and properties its parts. Petty's geography was anatomy expanded to a scale of miles.

Fighting off floodwaters, Irish rebels, and English lawyers, Petty managed to finish his survey as promised in 1656. In later years he analyzed the pile of information his team gathered and created a map of Ireland of unrivaled precision. Petty had found Ireland's true shape in a map more accurate than had ever been drawn before, without empty patches or coastlines trailing off into ignorance. Along the way, the boy had who survived for a week on a handful of walnuts made himself outrageously wealthy. The surveyor-general paid Petty £10,000, which would be only the beginnings of his fortune. He used a series of dubious trades, inflated prices, and an intimate knowledge of his own survey to acquire fifty thousand acres by the time he was thirty-three. The new science was not a business for monks with no interest in this world. Francis Bacon had promised from the start that it would let men control the world and profit from it, and Petty was one of the first to show how that profit could be earned.

Yet Petty also believed that he was still carrying on the scientific work he had done with his friends back in Oxford, extending his methods from cadavers to entire nations. With the survey he had hoped "to enlarge my trade of experiments from bodies to minds, from the motions of the one to the manners of the other, thereby to have understood passions as well as fermentations, and consequently to have been as pleasant a companion to my ingenious friends, as if such an intermission from physic had never been."

Petty was interested in souls, but not in order to save them. He saw souls as cogs in a nation's economic machine, capable of directing a body's labor in the production of wealth. Hobbes had figured out how the political machine could ensure peace, and Petty wanted to figure out how it could also make its citizens rich. He wrote a book he called *The Political Anatomy of Ireland,* in which he declared that both "the Body Natural and the Body Politic" have to be dissected. Trying to maintain a nation's economic health without understanding its political anatomy was as foolish as putting one's own health in the hands of "old-women and empirics." Petty admitted that his dissecting tools were far cruder than the ones he used on cadavers in Oxford, but they were a good start—"my rude approaches being enough to find whereabout the liver and spleen and lungs lie, through not to discern the lymphatic vessels, the plexus, choroidus, the volvuli of vessels within the testicles." Although Petty would later extend his method to England and elsewhere, he began with Ireland for the same reason that "students in medicine practice their enquiries upon cheap and common animals, and such whose actions they are best acquainted with."

Petty sought to weigh and measure all the parts of the body politic. By estimating how many people were in a country and how much labor they could carry out, Petty argued that it would be possible to engineer a country to run as prosperously as possible. He shared Hobbes's dislike of the clergy and estimated the minimum number of ministers a country needed, the rest to engage in more productive work. Petty estimated the productivity of nations (ten million Englishmen, he claimed, were worth more than thirteen million Frenchmen), he estimated their vice (from the number of people in jail), and he estimated their genius (from the number of difficult books sold each year).

Petty ultimately failed to get his grand accounting schemes established as the policy for Ireland, but along the way, he laid the foundation for statistics and economic planning that every government uses today. In the twenty-first century, we all fall under the knife of Hobbes's anatomist.

While Petty sought his fortune abroad, Willis remained in Oxford. Still struggling as a pisse-prophet, he remained loyal to his family, his fallen church, and his dead king. Every day his Anglican friends, including Christopher Wren, Ralph Bathurst, and John Fell, would come to his rooms to worship according to the illegal Book of Common Prayer. Around 1650 Willis moved out of Christ Church and into Beam Hall, a low-slung medieval house on Merton Street, where he continued to host the forbidden church. As many as three hundred people squeezed into the cramped rooms to worship and hear sermons from priests dressed in surplices.

Willis's church became known as the Loyal Assembly, "to which he sincerely adhered, even to the danger of his life," as an eighteenth-century biographer wrote. Three hundred notorious royalists crowding every day into the same small house were hard to keep secret, however. Fortunately for Willis, the new dean of Christ Church, a minister named John Owen, lived just a few doors away yet never once tried to break up the Loyal Assembly. Owen dressed like a dandy and wore enough powder in his hair to "discharge eight cannons," according to the Oxford diarist Anthony Wood, but his library was full of books by Galileo, Gassendi, Mersenne, Descartes, Bacon, and Harvey. He led troops on horseback with sword and pistol to crush royalist uprisings and wrote half a million words attacking the theology being taught in Willis's house. But he also opposed a national Presbyterian Church and the abolition of the Book of Common Prayer. Like Cromwell, Owen believed in allowing as much religious tolerance as politically possible.

Willis's private church helped keep the spirit of a royalist England alive for the next ten years, during "a peace more cruel than any war," as John Fell later wrote. The royalists who prayed in Beam Hall clutched their relics of the executed king, and the priests delivering the sermons did more than offer spiritual guidance. Willis's old classmate Richard Allestree, for instance, also worked as a spy and courier, slipping back and forth between England and Europe with secret instructions. The Loyal Assembly was a meeting place for royalists planning uprisings, writing propaganda against the Cromwell regime, raising money for the cause, and secretly ordaining priests.

For Willis, the services at his house not only sustained his faith but helped him form his philosophy of the soul as much as reading Gassendi or observing Anne Greene. Many Puritan sects held that the fate of our souls is foreordained. The elect are destined for heaven and discover truth not through human reason but through divine revelation. The Church of England, on the other hand, held that the fate of the soul remains open through life. Through reason one could come to understand and accept Jesus, but it was also obvious that illness and injury could easily cut life short and with it the time in which reason could guide a soul to God. For the Loyal Assembly, medicine therefore had a spiritual dimension, because it could give people enough life to make their salvation possible. Jesus, Willis wrote, believed that "the health of the soul should take its beginning from the restored health of the body."

Common illnesses such as typhus and malaria were not the only threats to salvation. Madness and other maladies could rob people of their reason as well. Each individual was like a tiny nation in need of a strong king to keep it in harmony with itself. In one of his sermons, Richard Allestree warned of the danger of passion, "which being by God and nature placed in a subserviency to reason, when it quits its proper station and assumes empire, it must needs disorder and subvert not only the state of Mind, but of everything upon which it has an influence." He compared passions dominating reason "as it fares sometimes with Magistrates in Popular insurrections," when the mind could be forced to approve terrible decisions—such as Charles I allowing the execution of his closest advisors—because the passions forced it to accept their fancies.

For Willis and the rest of the Loyal Assembly, England as a nation needed to be healed as much as its subjects. With the Civil War, Cromwell's rule, and the rise of Puritan fanatics, they had no doubt the nation was sick in soul and body. "How shall we pray for peace," Allestree asked, "that still retain our wickedness? This, O Lord, is our sorest disease. O Give us Medicines to heal this sickness, heal our souls, and then we know thou canst soon heal our land."

Sitting in a cramped room in Beam Hall surrounded by his fellow royalists, Willis listened to sermons like Allestree's and prayed passionately that England's soul would be cured. A decade later, he would found the science of neurology in his effort to heal it.

Chapter Six

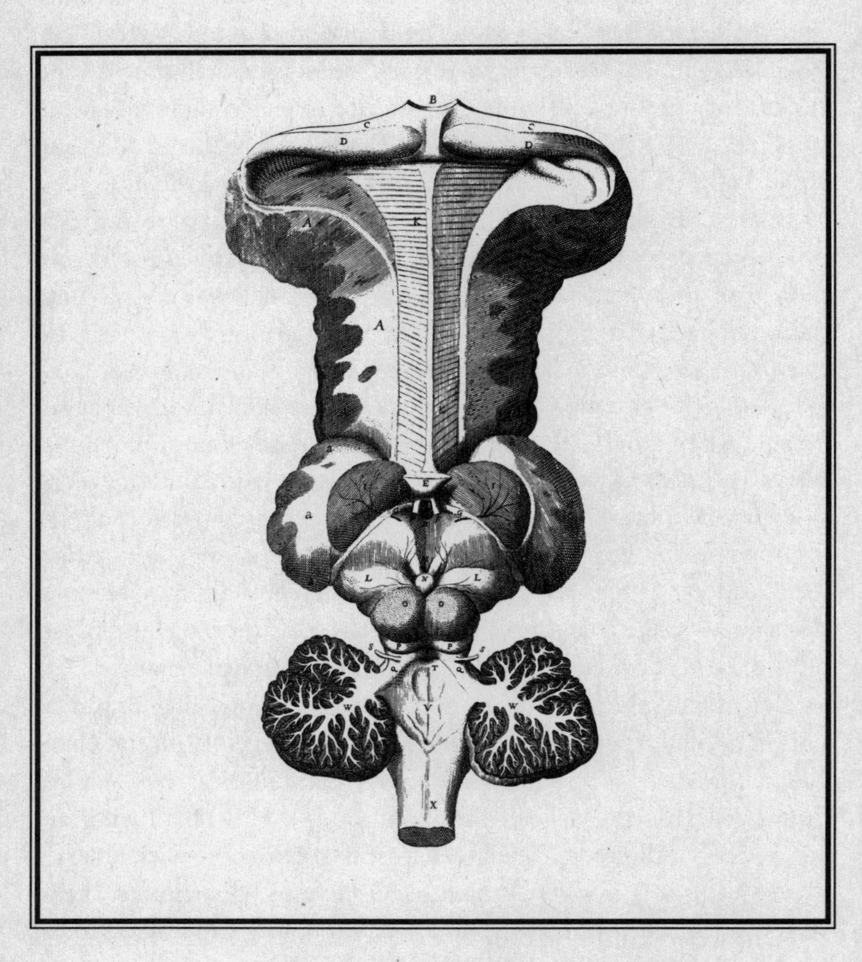

A PARTIALLY DISSECTED SHEEP BRAIN, FROM *THE ANATOMY OF THE BRAIN AND NERVES.*

The Circle of Willis

*I*n the years after William Harvey left Oxford, a rumor spread across Europe that he was dead. Harvey may have wished he was, but he would hang onto life for eleven more years. From Oxford he had taken refuge with his two brothers, both wealthy merchants who had also suffered for their allegiance to the king. After Charles I was captured, Harvey managed to attend the king only once before his execution. Parliament classified Harvey as a delinquent, fined him £2,000, and forbade him to come within six miles of London without its permission. Harvey's medical practice had dwindled to a few acquaintances, and so he spent most of his time with his brothers, drinking coffee and dissecting animals. He battled his gout by sitting outdoors on frosty days with his foot in a bucket of water. John Aubrey recalled "a pretty young wench" who Harvey had wait on him, "which I guess he made use of for warmeth-sake as King David did, and took care of her in his Will."

Harvey was visited in his seclusion in 1648 by a London doctor named George Ent. In 1641, Ent was the first Englishman to defend Harvey in print, and seven years later, Harvey was still being attacked by critics across Europe. When Ent finally met Harvey, the old man was studying "Nature's Works, and with a countenance as cheerful, as Mind imperturbed, Democritus like, deeply searching into the causes of natural things," as Ent later wrote.

"Are all affairs well, and right?" Ent asked him.

"How can they be," Harvey answered gravely, "when the Commonwealth is surrounded with intestine troubles, and I myself as yet far from land, tossed in that tempestuous ocean?" If not for his studies and his memories, Harvey said, "I know not what could prevail upon me to desire to survive the present."

Ent told Harvey that his admirers "greedily expect the communication of your further experiments."

Harvey smiled. "And would you then advise me to quit the tranquility of this Haven, wherein I now calmly spend my days: and again commit myself to the unfaithful Ocean?"

Ent pressed Harvey further. Finally, the old doctor hobbled away to dig through his papers and books and handed Ent a manuscript, one of the few works of his that had escaped destruction during the war. In it Harvey had summed up his work and ideas about embryos over his entire career, from the royal deer of Charles I to the chicken eggs of Oxford. Harvey had made no effort to publish it and might never have if not for Ent, who offered to handle all the work of getting it to a press—"the mere office of a Midwife." Harvey allowed him to take the manuscript away. Leaving Harvey, Ent felt like "another Jason, enriched with the Golden Fleece."

Disputations Touching the Generation of Animals was published in 1651. In it, Harvey explained that all life begins as eggs, that humans and deer come from eggs just as chickens and frogs do, despite the fact that he had never seen the microscopically small eggs of mammals. *Ex ovo, omnia* was his motto. Many natural philosophers at the time believed that an animal's anatomy was already sketched out in miniature form in an egg, but Harvey took the opposite position: the organs of the body come into existence only gradually, taking shape over time. Only a soul could guide the egg's development to its destined form. He scoffed at natural philosophers who claimed that blind atoms could do the job, passively following the laws of physics. In Harvey's view, these mechanists insulted the workmanship of God.

Seeing his new book bob out onto the impassive, faithless sea of public opinion did not lift Harvey's spirits. In *Disputations Touching the Generation of Animals,* he promised that he would write about "the soul and its affections, and how art, memory, and experience are to be regarded as the conceptions of the brain alone," but he never published anything on the subject. Harvey mourned the loss of the old England and scorned the chaotic society that had replaced it. He had considered endowing a professor's chair at Cambridge University but abandoned the idea, complaining that "I should do nothing other than make Anabaptists, Fanatics, and all manner of thieves and parricides my heirs." In the year *Disputations Touching the Generation of Animals* was published, Harvey tried to commit suicide with laudanum. It did not kill him but did clear his kidney stones. With his health restored (if not his spirits), Harvey survived for another five years, "till he doted as I have heard say, and would talk but very weakly," according to the diarist John Ward. In 1657, at age seventy-nine, Harvey expressed his weariness to a friend: "It seems to me indeed that I am entitled to an honourable discharge." That year, he finally got it.

As hard as it was for Harvey to survive so long, he managed to enjoy a rare privilege: he saw his ideas about the circulation of blood—and his new way of doing science in general—finally begin to take hold. Hobbes wrote that Harvey "is the only man, perhaps, that ever lived to see his own doctrine established in his life-time." As Charles's execution receded into the past and Harvey grew old, Parliament no longer saw him as a dangerous royalist. Harvey even financed a magnificent library for the College of Physicians, which erected a bust of him made of white marble and inscribed with praise for his discovery of the motion of blood and the origin of animals.

In Oxford in the mid-1650s, it felt as if Harvey had never left. Thomas Willis, Christopher Wren, and the other members of their remarkable circle carried on Harvey's research on circulation. Despite years of bloodshed, chaos, sickness, and anxiety, they managed to carry out experiments and perform dissections. They posed questions instead of demonstrating conventional wisdom. They measured and drew, hypothesized and tested, tackling the questions Harvey had left unanswered after clearing away many of Galen's claims. If the heart drove the blood in a circle through the body, what did that mean for the cause of diseases? What was a fever? If the heart did not beat in time with the lungs, then what was the purpose of breathing? Could the flow of blood somehow be harnessed by physicians? Harvey died just as the Oxford virtuosi were beginning to answer these questions. In the process, they were laying the groundwork for an even more ambitious project that would begin four years after Harvey died: to discover the workings of the brain.

—

During the mid-1650s Thomas Willis spent a lot of his time dealing with the preoccupation of every physician of his age: fever. According to Willis, fever was "a disease by which the third part of mortals have still fallen to this day," yet it was for the most part a mystery. Some kinds of fever came regularly with the summer, while others came with droughts or wars. Some were intermittent, some continual, some struck only women in childbirth, others mainly children. Some made their victims flow with phlegm, and others left them with an endless thirst. But fevers all had in common a raging heat and the power to kill, sometimes claiming thousands in a matter of weeks.

Fevers had killed Willis's parents during the Civil War, and he had watched epidemics regularly hack their way through Oxford. As a doctor, he became even more obsessed by fever, keeping detailed records not just of individual cases but of overall patterns—of the ebb and flow of fevers through the year, of the different outcomes they had in different sorts of people.

Willis did not name the fevers he studied as we do today—malaria, encephalitis, and the like—but he could distinguish between them with unprecedented skill. In one year, for example, he carefully recorded how "about the end of April, suddenly distemper arose, as if sent by some blast of the stars." In the space of a week, over a thousand people fell sick around Oxford, coughing, spitting, with mucous dripping from their nose and down their throat, suffering as well from a lack of appetite, weariness, and "a grievous pain in the back and limbs." Willis described the course of this new kind of fever, the way some people could go about their business (with plenty of complaining) while others were wrecked and sometimes died. In the process, Willis wrote the earliest known clinical description of influenza.

Fevers took up so much of Willis's time that he had plenty of opportunity to mull over their causes. "Sitting oftentimes by the sick," he wrote, "I was wont carefully to search out their cases, to weigh all the symptoms, and to put them, with exact diaries of the diseases, into writing, then diligently to meditate on these, and to compare some with others, and then began to adapt general notions from particular events." Willis decided that it was time for a new explanation of fevers, even if it might seem, as he later wrote, "as if anyone should go about to describe the midst of our country for a land before unknown."

It was becoming clear that the body was not the country it had once seemed. "Those things have not pleased the men of our Age, which did those of the former," Willis declared, "because the ancients relying on a false position concerning the motion of the blood, proceeding as it were through slippery and moist places, often fell foully and dangerously." In discovering the circulation of the blood, Harvey had created, in Willis's words, "a new foundation of medicine."

Without Galen's humors, Willis had to look for some new way to account for fevers. To find an answer, he would have to step back from the narrow question of fevers and ask how a healthy body kept warm. He would have to bring together his work on the dissecting table with his work in the alchemist's laboratory. Galen had claimed that diseases were caused by humors massing up in certain regions in the body. But that was absurd, since the heart drove the blood in a circle. There were no humors, Willis decided, but only blood and its excrements.

—

In the years that Willis spent looking for the cause of fevers, he hired a crook-backed assistant named Robert Hooke. Hooke, like his friend Christopher Wren, came to Oxford from a royalist family wrecked by the war. His father, a curate on the Isle of Wight, had been fiercely dedicated to the king and Archbishop Laud. After the war Parliament laid heavy fines on him for his loyalty, and the disgrace appears to have been too much for him. He died under mysterious circumstances in 1648 when his son Robert was only thirteen. One hint that he may have committed suicide is the careful arrangements he made for his son in his will. He left forty pounds, his best joined chest, and all his books to Robert, who used his inheritance to make his way to London. There he enrolled at Westminster School, where his genius shimmered as brightly as Wren's. He was the sort of fifteen-year-old boy who could master all six books of Euclid in a single week. Willis took in this royalist orphan when he arrived at the university, and he quickly recognized the boy's gifts. Soon Hooke was part of the Oxford circle, working with Wren on new telescopes and microscopes, inventing springs for pocket watches, and building flying machines for Wilkins.

In some ways, Hooke, Wilkins, and the rest of the Oxford circle were bringing Bacon's utopia to life in miniature, creating a peaceful space in England where the new science could be pursued by a network of friends. But the circle's existence hung in a balance that was more delicate than its members would admit. They were part of a scientific revolution that was determined to replace Aristotle and Galen with Galileo, van Helmont, and Harvey, but they did not want to be mistaken for atheists using the new science to dismiss God from the world. Nor did they want to sweep the universities away and replace them with nothing more than the fanatical light of divine inspiration.

In the early 1650s, this middle ground was guaranteed to satisfy almost no one. The old guard was appalled by the newfangled notions that threatened to replace the ancients. "I pity to see so many young heads still gaping like chameleons for knowledge, and are never filled, because they feed upon airy and empty phansies," one critic wrote. At the same time the Oxford circle came under attack from Puritans. Puritans had no quarrel with the new science per se—many believed that it promised to turn England into a new Eden—but the Oxford circle seemed too easily distracted from the divine mission at hand. They seemed to be having too much fun, perhaps. One Puritan critic called them "mere Moral men without the power of Godliness." Some found their obsession with microscopes to be unseemly—they "were good at two things, at diminishing a Commonwealth and at Multiplying a Louse."

For some radicals, the Oxford circle was an irrelevant distraction from their quest to destroy the universities. An influential minister named William Dell denounced life at the universities as a "daily converse with the Heathens, their vain Philosophers, and filthy and obscene Poets." He demanded that Oxford and Cambridge be abandoned and new universities built in every English city, where students would learn useful trades.

Another radical minister, John Webster, launched an attack of his own. "If we narrowly take a survey of the whole body of their Scholastick Theologie," he declared, "what is it else but a confused Chaos, of needless, frivolous, fruitless, trivial, vain, curious, impertinent, knotty, ungodly, irreligious, thorny, and hell-hatch't disputes, altercations, doubts, questions, and endless janglings, multiplied and spawned forth even to monstrosity and nauseousness?"

Webster had studied Paracelsus and had served as both surgeon and chaplain in the parliamentary army during the war. He believed theology had been rendered obsolete, because the Puritans had shown that reason could never bring man the gift of grace. Webster accused Oxford not only of preparing bad ministers but also of teaching bad science. He claimed that it still taught Ptolemy instead of Copernicus, that it fawned on Galen—that "ignorant pagan"—instead of van Helmont and the "most admirable and ravishing knowledge" of Paracelsus. Alchemy alone could reveal how nature worked, Webster declared, but it was missing from Oxford. He demanded that students put down their Aristotle and march into a laboratory. "I dare truly and boldly say," he wrote, "that one year's exercise therein to ingenious spirits, under able Masters, will produce more real and true fruit, than the studying Aristotelian philosophy hath brought forth in many centuries."

One might expect the Oxford circle to have welcomed Webster as an ally rather than an enemy. After all, Willis and Bathurst were in their laboratory, following in the footsteps of van Helmont and Paracelsus. Seth Ward and John Wilkins were both champions of Copernicus. But the Oxford circle looked on Webster, like Dell, as a fanatic. Both had abandoned reason, and their mystical instincts left them with muddled ideas about science. Yet the Oxford circle could not simply ignore their attacks. The times were too dangerous for that.

Ever since the end of the Civil War, Parliament had been promising to overhaul Oxford and Cambridge, but until now it had been distracted by war. With England finally at peace, Parliament began to look inward, at how England itself should be reworked. Members debated what sort of national church, if any, should be established, and what shape the Commonwealth's government should take. The debates dragged on for months, and Cromwell and the army became impatient. Parliament was more interested in consolidating its own power and making money from seized royalist land than in building a new nation. Cromwell began to think the only way to deliver England would be to make himself king. By the beginning of 1653, the army was sending threatening addresses to Parliament. Cromwell tried to negotiate between the two sides, but by the spring his own patience was spent. He marched into a session of Parliament and declared, "I say you are no Parliament." He summoned a few dozen soldiers into the House of Commons and had them drag out the Speaker along with his mace. The republic was dead.

Cromwell set out to rebuild the government by having 140 men selected from across the country. They gathered to form a body he called the Assembly of Saints. "You are at the edge of the promises and prophesies," he told them. England's radicals took Cromwell at his word and swarmed into London, urging the Assembly of Saints to turn England truly upside down. Levellers returned from exile, calling for a democracy and stirring up unrest in the army. John Webster came with other ministers to London to preach in the streets. Some shouted that the saints were preparing the way for the monarchy of Jesus. The members of the Assembly of Saints paid attention, and they demanded that the church give up its tithes and that Oxford and Cambridge Universities be stripped of their land and wealth.

The Oxford circle defended itself against these attacks with a pamphlet by Wilkins and Ward. The two men pointed out the contradictions of Webster and Dell, even supplying an easy-to-read chart showing all the places where Webster had plagiarized Gassendi. Far from being an Aristotelian indoctrination camp, they claimed, Oxford was immersed in the new science. They pointed to Wren's and Hooke's work with microscopes, which they were sure would uncover life's smallest, most important ingredients. "When the microscope be brought to the highest, whence it is apace arriving, we shall be able to give the seminal figures of things which regulate them in their production and growth," they wrote.

They proudly presented Willis and the rest of their "Chymicall society" hard at work in the service of natural philosophy. As for medicine, they boasted about Petty's and Willis's resurrection of Anne Greene. "Surgery as well as Physick hath even in our time been extremely advanced, this place hath given late instances of both; (particularly in recovering the Wench after she had been hanged at least half an hour, and others which I could mention)." Theology and natural philosophy were both in good hands at Oxford, Wilkins and Webster assured their countrymen, but if

England followed the advice of Dell and Webster, ignorance would spread through the country until "the Romans will come and take away their Place and Nation."

The Oxford circle won a temporary victory. The Assembly of Saints did not pursue the destruction of the universities. By the end of 1653, Cromwell regretted having created the Assembly at all. It was too radical for his taste, too reckless with the rule of the country. His allies staged a mass abdication from Westminster Hall, and Cromwell shut down the Assembly for good. In its place, he set up a new constitution and made himself Lord Protector. Parliament would still meet, but only rarely, and it would be elected only by men who owned property to the value of at least £200. Cromwell would rely on a council of his closest allies to run the country. He would not allow radicals to overturn England. Under his Protectorate, the universities would be safe.

The attacks were hardly over, though. In 1654, a sixteen-year-old girl named Elizabeth Fletcher walked naked through the streets of Oxford. She belonged to a new sect that called itself the Society of Friends, which had been gaining strength in the north of England for several years and was now sweeping across the rest of the country. Members of the Society believed that each person is his or her own priest and that true religion comes not from theology but from an inner light that can overwhelm a soul and make it quake—hence their nickname, Quakers. The defeated political radicals found comfort in new sects like the Quakers, and under Cromwell's policy of religious toleration, the sects blossomed.

Quakers believed that their fellow Englishmen were sunk in a hypocritical stupor and the only way to rouse them was by shock—the sort of shock Elizabeth Fletcher hoped for when she walked naked through Oxford. Other Quakers burned Bibles and stood up in the middle of church services to shout down ministers. They treated all men as their equals, even refusing to tip their hat to their so-called superiors. They preached in Oxford's marketplace and charged into the colleges to declare the word of the Lord.

The Quakers proved to be hugely popular, particularly in places like the slums of Oxford. Oxford's scholars responded by having them thrown into jail and whipped. Elizabeth Fletcher was attacked and nearly killed. Convinced that the Quakers were insane fanatics, the scholars tried to shock them out of their delusions. Some of them broke into a Quaker meeting house during a service, hauling in squealing hogs and raving madmen. They stormed through the meeting, cutting off half the beard of one man, sticking pins into worshippers, riding their backs like mules, barking and meowing.

As one minister preached in 1653, the visions of fanatics "are nothing but the distempers of a disaffected brain," but the Quakers were perfectly happy with the fanatic label. "Men therefore must now be sober to God," one Quaker declared, "but stark mad with the church, in plaguing, vexing and destroying all her delicacies."

—

With naked, railing Quakers wandering through their town and Puritan ministers calling for the university's destruction, the Oxford circle found much to worry about. Of all their enemies, however, they felt most threatened by that feeble old exile, Thomas Hobbes.

When Hobbes and Petty had parted ways in Paris in 1646, the philosopher was the darling of the exiled royalists. Hobbes wrote a book called *De cive* (*On the Citizen*) in which he argued that his peculiar philosophy revealed how to govern a nation properly. Like so many other Englishmen, Hobbes wished for some vaccination against future civil wars and was confident that his philosophy "will not only show us the highway to peace, but will also teach how to avoid the close, dark, and dangerous by-paths of faction and sedition." For Hobbes, absolute rule was the only true protection.

A few hundred copies of *De cive* circulated among the royalists in Europe, who saw it both as a veiled endorsement of the king and as an attack on Parliament. It was wise to stay on the good side of the exiled royalists, who were known to stalk Parliament's friends on the Continent and assassinate them. But Hobbes did not stay on their good side for long, thanks to a new book he started shortly after Charles I was beheaded, which was to combine all his work on physics, anatomy, psychology, and politics into a single staggering synthesis. "He walked much and contemplated," wrote his friend John Aubrey, "and he had in the head of his staff a pen and ink horn, carried always a notebook in his pocket, and as soon as a thought darted, he presently entered it into his book or otherwise he might have lost it."

The universe, Hobbes wrote in his notebook, was made only of matter. Our bodies were no exception, nor were our thoughts. Echoing the mortalist Richard Overton, Hobbes declared that the soul survived death no more than the taste of bread survives a loaf. Reason was not a candle of God but simply the sign of an ordered mind that could put proper names to its thoughts. Given the nature of men—their fear of death, their hunger for life, their struggle for power—Hobbes argued that only one kind of government would do: an absolute ruler to whom the people surrendered their natural liberty. Laws did not come from God, but were simply the means by which people could live together peacefully under their ruler, a "mortal god," as Hobbes called him.

With confidence born out of his knowledge of the absolutes of geometry, Hobbes declared that in order to preserve a happy nation, its ruler must dictate its religion. God was simply the world's cause, and "power of this kind necessarily elicits worship." A country's citizens had to accept whatever kind of worship their ruler imposed. Hobbes did not bother much about exactly which religion a sovereign ought to choose, but warned that if he lacked complete control over religion, delusion and strife would follow. He mocked both the Catholic and Protestant churches for dealing in the business of fairies and ghosts and worried that the promise of eternal life and eternal torment after death would undermine the sovereign's authority.

Hobbes published *Leviathan* in 1651 and considered it an eminently reasonable, respectable piece of work. He sent the book out to acquaintances whose intelligence he respected. At Oxford, Ralph Bathurst and Seth Ward got copies. When Prince Charles escaped from England after his defeat, Hobbes personally gave his old student one as well.

But *Leviathan* stirred up a public rage as few books have ever done. Hobbes believed that whoever happened to be in power should be obeyed, and his book expressed no particular preference for who the absolute ruler should be. To the royalist court in France, *Leviathan* felt like a surrender to Cromwell. It offended them with its dismissal of the immaterial soul as they grieved for their beheaded king, as they clutched cloths soaked in his blood that they believed could heal.

By dismissing Rome, Hobbes earned the ire of the powerful English Catholics in exile with the prince. By mocking the Church of England, he lost the confidence of Protestant royalists, who believed he was comparing the church's authority to "the kingdom of fairies." Old friends became new enemies. It took only a few months for them to get Hobbes banished from the exile court. At age sixty-three, struggling now with a palsy probably caused by Parkinson's disease, Hobbes realized that he now had nowhere to go but back to England. "I returned to my homeland," he later wrote, "not quite sure of my safety. But in no other place could I have been safer. It was cold; there was deep snow; I was an old man; and the wind was bitter. My bucking horse and the rough road gave me trouble."

Hobbes arrived in England in 1651 and managed to settle into a life of relative comfort in London. He renewed his friendship with William Harvey (the two had known each other for decades, and Hobbes had even watched Harvey's dissections of King Charles's deer). Hobbes also buzzed like a gadfly through the intellectual circles of London. The astronomer Seth Ward recalled that "if anyone objected against his dictates, he would leave the company in a passion, saying, his business was to teach, not dispute." *Leviathan* brought Hobbes fresh trouble in London, where booksellers complained that the book was dangerous to religion. Within a year, pamphlets appeared condemning Hobbes. In *Leviathan Drawn Out with a Hook,* Alexander Ross declared that Hobbes "vomited up the condemned opinions of the old heretics."

Before long Hobbes was under attack by the Oxford circle as well. He had condemned England's universities for promoting old superstition instead of real knowledge and said that the best remedy would be to make *Leviathan* required reading for all their students. At first, the Oxford circle's response to his attack was mild. Hobbes's earlier books had much to admire, they said, but he had lurched off course with *Leviathan.* Seth Ward declared that Oxford had changed a lot since the old man had been a student there. Hobbes took the remark as an insult and afterward refused to be in the same room as Ward.

The Oxford circle couldn't help but notice that all around them students were reading *Leviathan,* as Hobbes had recommended, and some tutors preferred to use Hobbes in their lessons rather than Cicero or Aristotle. If Hobbes's beliefs about the soul and God spread too far, they could spell disaster. Hobbes's writings about religion, Ward wrote, made his limbs "numb with horror and indignation." Relations went from strained to vicious. Ward insinuated that Hobbes was a plagiarist and a bad mathematician. Hobbes countered by showing off his geometrical gifts, demonstrating that he could draw a circle and a square of the same area using only a straight-edge and a compass. John Wallis, the Oxford mathematician, took every opportunity to attack Hobbes's math and his religion, writing that "our Leviathan . . . is attacking all religion." Wallis hoped that by destroying Hobbes's reputation as a mathematician, he could tear down his philosophy altogether.

Wallis and Hobbes would wage a pamphlet war for twenty years, and other members of the Oxford circle would keep up the fight as well. Thomas Hobbes had done something both unthinkable and dangerous. He had taken the mechanical philosophy to a logical extreme, doing away with everything beyond the world of matter. The Oxford circle mixed their mechanical philosophy with gentle skepticism and an abiding faith in the spirit world—and certainly did not want to be confused with the likes of Hobbes.

Hobbes would survive their attacks. With *Leviathan,* he had essentially invented political science. Since 1651, every theory for governing a nation has had to confront Hobbes in one way or another. He built his political science out of decades of sharp observation of his fellow Englishmen at their best and their worst. His influence as a political thinker grew even though he believed that his philosophy sprang naturally from the organization of the human body and even though his knowledge of that organization was simply wretched. Just as Descartes was supremely confident in the pineal gland as the cockpit of the soul, Hobbes was sure that the heart was the center of sensation. The mind was "matter in motion" to Hobbes, the brain a hazy globe filled with spirits on their way to other organs. If the modern science of the brain had been founded on Hobbes's ideas, it would have stalled before it had even started.

—

Hobbes and the other enemies of the Oxford circle exposed a serious weakness. These virtuosi investigated a staggering range of questions but had yet to present any coherent account of what they were doing. It was fine to map the moon and make invisible ink, but the virtuosi of Oxford sometimes seemed to be dabbling without purpose, just as England's aristocrats had been amusing themselves with scientific pastimes for over a century. The Oxford circle might sing the praises of Bacon and Harvey in the back of an apothecary shop, but these natural philosophers embodied contradictory views of natural philosophy. Bacon wanted to gather facts rather than let old dogmas tell him what to think. Harvey was guided by Aristotle (whom Bacon called a dictator), but Harvey also carried out experiments to test his assertions. The Oxford circle badly needed an anchor and a spokesman. In 1655, they found both in a gaunt, sickly young aristocrat named Robert Boyle.

Most members of the Oxford circle fell into one of a few categories. There were young royalists diverted from the church to medicine, such as Willis and Bathurst, and broad-minded Puritans from London, such as Wilkins and Wallis. Robert Boyle was neither. He was born in 1627 to Richard Boyle, the richest subject of Charles I. His father had traveled to Ireland in the late 1500s to make his fortune. With faked letters of introduction, he had charmed his way into powerful English circles and was eventually appointed to check the legitimacy of all the land titles that came into the Crown's possession. Using this inside information—and the help of powerful friends—he bought cheap land and made shady deals, buying up much of southern Ireland within a few years. King James gave him the title Earl of Cork, and for more than forty years he controlled his estates like a private kingdom, building fifteen castles to protect them from Irish rebels and filling them with a personal army. Boyle founded towns for his tenants, built harbors and bridges, stripped lands of most of their forests, and dug their mines bare. He made his motto "God's providence is my inheritance" and his life a continual search for power and wealth. He kept scrupulous journals for decades, noting every transaction he made down to the shilling, but never once did he mention reading a book. He fathered fifteen children, married off the daughters badly, and watched his sons for the most part turn into drunken, violent louts. It was not the sort of family to produce a natural philosopher.

From the start, Robert proved a peculiar Boyle. As a boy, he read omnivorously and wrote endlessly. Sent off on a grand tour of Europe with his older brother Francis, Robert picked up languages like a magnet and blended into every country he visited. In Italy he read Galileo for the first time, and in Padua he saw arteries, veins, and nerves preserved and laid out on boards.

But the journey was most important to Boyle for a religious awakening he had at his tutor's home in Geneva. He was awakened one night by booms of thunder and flashes of lightning that came so fast that he thought the Apocalypse was coming, that the Earth was going to be swallowed by fire. The wind and the rain that followed were so fierce that Robert was sure the Day of Judgment was at hand. Suddenly he felt utterly unprepared for that moment and prayed that if the end of the world did not arrive that night he would live a moral life. When the morning came with clear, quiet skies, Robert was sure he had been given a sign.

For the rest of his life, Boyle would chase after grace and be tormented by a feeling that he was unworthy of God's love. He would relentlessly question whether he was pious enough, if he had actually committed some evil that he was unaware of. His uncertainty turned his life into an endless round of questions, a perpetual suspicion that he had not yet found Truth.

In 1641, after two years abroad, Robert and Francis were ready to come home from their tour, but at that moment their privileged lives were suddenly interrupted. The Irish rebelled against their overlords, including the earl of Cork. The earl counterattacked with an army of his tenants and saw his eldest son killed in battle. The rebels plundered his estates and drove him into one of his castles. In a desperate letter, the earl promised his sons £250 to pay for their passage from Geneva to a safe port in Ireland or to Holland, where the boys could fight with the prince of Orange against Catholic Spain. "They must henceforward maintain themselves by such entertainments as they get in the wars," he wrote. On no account, their father warned, must his sons come back to England. The earl could not stand the humiliation of the boys being seen as paupers. When Charles I made a truce with the Irish in 1643, the earl simply died in despair.

The money the earl promised his boys never came. Francis managed to make his way back to Ireland to fight the rebels, but Robert, young, suddenly poor, and in bad health, stayed behind in Geneva with their tutor for two years until his money ran out. He decided to get to England somehow, although he wasn't sure what he'd do when he got there. His tutor lent him enough money to buy a few jewels, which he sold along his way to London.

Boyle arrived in London in 1644 without any idea of what had become of his family. By accident, he learned that his favorite sister, Katherine, had moved to London as well. Thirteen years older than Robert, she had married Lord Ranelagh, a drunken oaf described by one of her friends as "the foulest churl in Christendom, whose best point was that he was nightly dead drunk and so probably not quarrelsome." When the Irish rebellion broke out, Katherine had been besieged in one of Ranelagh's castles. After two years, she found safe passage to London, and when she decided to stay there, Lord Ranelagh offered no quarrel. They never saw each other again.

Most of the Boyles were royalists, but in London Lady Ranelagh became friends with reform-minded Puritans and members of Parliament. John Milton tutored her son. At her house, she hosted talks on alchemy, the reform of medicine, the new science, and religion. She introduced her brother to her friends, who immediately treated the boy as their intellectual equal. Thanks to his sister's connections with Parliament, Robert was able to get back some of the family's seized wealth, including a manor in southeast England.

Boyle settled into the decrepit house while the Civil War still raged. The forests around the manor were full of soldiers turned thieves profiting from the chaos. Boyle himself was not trusted by either side. Royalists saw him as the brother of the famous Lady Ranelagh, a powerful member of London Puritan society, while Puritans were suspicious of Boyle's royalist father and brothers. Once they went as far as arresting him, but their suspicions never came to anything.

Secluded in his manor house, Boyle continued his erratic, obsessive self-education. To find the true meaning of the Bible, he taught himself Chaldean and Syriac. He wrote long, sanctimonious essays on morality. But three years after he moved to his new home, Boyle had a second conversion, this time to science.

One of Boyle's neighbors, Nathaniel Highmore, had been a protégé of William Harvey during the siege of Oxford five years earlier. Highmore welcomed Boyle to his house, where the young man could watch his grisly experiments on live dogs, opening their chests to study how they breathed. There Boyle looked through a microscope and was dazzled by the fine hairs on the leg of a cheese mite. Meanwhile, the alchemists in Lady Ranelagh's circle initiated Boyle into the chase for the philosopher's stone. Boyle had a laboratory set up at his manor, complete with a massive earthen furnace, alembics, compounds, and a staff of assistants to carry out the work. "Vulcan has so transported and bewitched me," he wrote to his sister, "that the delights I taste in it make me fancy my laboratory a kind of Elysium."

Alchemy seduced Boyle on many levels. The alchemists he worked with believed that with the rising fortunes of the Puritans, England was now ready to cast off Galen and embrace a new, chemical medicine based on the ideas of Paracelsus and van Helmont. Boyle believed in their campaign, and he also hoped to find cures for himself. Since childhood he had suffered from fevers and weak eyes, and now he developed kidney stones that would torment him for the rest of his life. As a boy, he had once nearly died from medicine he had been given for diarrhea and had become determined to take only his own remedies. Each morning Boyle checked his weathervane so that he could take a medicine to defend himself against the direction of the wind that day.

Boyle also believed that alchemy could become a religious mission. By understanding the natural world, one could better appreciate God's handiwork. The philosopher's stone might make it possible, he thought, to talk to angels. By focusing on what everyone could see in nature, alchemy might even heal England, which Boyle worried was flying apart into hundreds of sects that to him seemed to have little interest in Jesus. "This multiplicity of religions," he said, "will end in none at all."

Boyle's hopes for alchemy made him impatient with some of the alchemists he met. Some seemed to be interested only in finding recipes for tinctures or purifying precious metals and were completely indifferent about what their work might reveal about the hidden workings of the world. Others were too eager to jump from a few discoveries to wild, all-encompassing theories. They carelessly handed out names to substances, creating confusion and blocking progress. Boyle wanted to question nature relentlessly, the way he questioned his conscience, and to do so through experiments.

There were some alchemists whom Boyle respected. They were his first teachers, helping to turn him into a true scientist. The one who influenced Boyle the most was a fiery young American named George Starkey.

As a student at Harvard University, Starkey had become interested in alchemy and immersed himself in the work of van Helmont. Soon he had created a substance he claimed was the alkahest—a universal solvent that was supposed to turn all things to water, to the original essence. Starkey had become convinced that he was also close to finding the philosopher's stone but had grown frustrated working with the shoddy furnaces and glassware made in New England. To get the equipment he needed, he had immigrated to London in 1650.

Starkey needed to attract the attention of powerful, wealthy patrons in London such as Boyle, which he did by inventing an alter ego. He claimed that he had learned everything he knew about alchemy from an adept (a master of the art) he had known in New England. This mysterious man could make a withered peach tree grow new fruit and an old woman sprout new teeth. His teacher wandered as an outcast in America but had given Starkey some manuscripts to be published. When Starkey published them under the pseudonym Eirenaeus Philalethes, they became hugely popular. It was rumored that the American adept was a herald for the prophet-alchemist that Paracelsus had predicted was coming. Eirenaeus himself promised that soon gold would be as cheap as dung. Since Starkey was the only person who seemed to have contact with Eirenaeus, he instantly became the darling of London's alchemists.

With his new patrons, Starkey hatched schemes to make perfumes and artificial diamonds, even to produce ice in summer. With Robert Boyle, he worked on medicines, trying, among other things, to create essence of copper, a substance that was supposed to be a powerful healing drug. The fumes in his badly ventilated lab nearly killed Starkey in the process, but as reckless as Starkey was, he was also a rigorous experimenter. Starkey kept careful notes of his procedures, which were precise enough to let a reader reproduce them. Starkey got little credit for bringing this discipline to science. He wrote a recipe for an alchemical compound known as philosophical mercury that was so detailed and explicit that historians later decided that a fraudulent alchemist like Starkey could not have authored it. They gave the credit instead to someone who had simply copied the recipe years later: Isaac Newton.

For Starkey, one of the most important mysteries of nature was what matter was made of. Throughout the Middle Ages, alchemists had kept the heresy of atoms alive. Some of them held that Aristotle's fire, air, water, and earth existed as invisible particles that could combine into other forms of matter. The philosopher's stone transmuted base metals as its tiny particles invaded their pores and converted them into gold. Starkey inherited the tradition, arguing among other things that materials had different densities because the atoms that made them up were packed tightly or loosely.

In England in the 1600s, Epicurus's theory of atoms still faced contempt. "Let that beastly Epicure's mouth be now sealed up in dumb silence," one Aristotelian wrote. But as Boyle learned about atoms from Starkey and other alchemists, he came to believe they were real. They allowed him to understand what happened not only inside his laboratory but in the outside world as well—why, for example, the invisible scent of a partridge could be smelled by a setter days later. Boyle granted that atheists might be attracted to atoms, but he did not believe that atoms shut God out of the world. When Boyle looked down through a microscope at the delicate joints of a fly's legs, he marveled at "what a multitude of Atoms must concur to constitute the several parts external and internal necessary to make out this little engine." Even at the finest scales of nature, God's handiwork was visible, and what better material for Him to work with than atoms? Boyle found more confidence in the books of Pierre Gassendi, who had explained how atoms could be squared with Christianity, how God exercised His will by arranging the world's atoms at the beginning of Creation so that they would wander through the universe in a way that ultimately obeyed His will.

Still, Boyle said little publicly about atoms or about alchemy. He was torn by his desire for peace and stability and his desire for a revolution in knowledge. He worried that revealing the deepest secrets of alchemy would "much disorder the affairs of mankind, favour tyranny, and bring a general confusion, turning the world topsy turvy." Boyle had seen kings pulled down and didn't want to bring down another with his secrets.

In 1652 Boyle left his laboratory, setting out on a long dreary trip to Ireland. His mission was to negotiate with the government for the return of some of the land his father had lost in the bloody 1640s. The legal drudgery lasted for two years, but it was worth the effort. Boyle ended up with a lavish income of almost £3,000 a year. Away from his laboratory, Boyle grew bored until a friend suggested he get in touch with another virtuoso in Ireland, William Petty, who had arrived from Oxford a few months earlier. Petty was then in the midst of his survey, which would let him build a little Irish empire as Boyle's father had thirty years before. Boyle turned Petty into yet another unofficial professor, learning from him how to dissect animals and how to see for himself the newest discoveries in anatomy. In those hours he spent "conversing with dead and stinking Carkases," Boyle wrote, he had delighted "in tracing in those forsaken mansions, the inimitable Workmanship of the Omniscient Architect."

Boyle was now twenty-seven years old, and it was becoming clear that he was not going to live the life of a normal aristocrat. He hated to "squander away a whole afternoon in tattling of this Ladys Face & tother Lady's Clothes; of this Lords being Drunke & that Lord's Clap; in telling how this Gentleman's horse outrun that other's Mare." Although he is known to have courted at least one woman, he showed no interest in marriage—an indifference that may have been bred by all the marital disasters in his family. "Being born heir to a great family is but a glittering kind of slavery," he once said, because it "is ever an impediment to the knowledge of many retired truths, that cannot be attained without familiarity with meaner persons." Boyle wanted only to turn alchemy into true natural philosophy, and the only way he could do that, he realized, was to move his operations to Oxford.

For years Boyle had heard the stories about how Oxford was rising from the ruins of war. Thomas Willis sent him accounts every week of his own chemical experiments and impressed Boyle as "a very ingenious experimenting gentleman." Boyle liked the way John Wilkins had created a safe haven for natural philosophy, where political and religious conflict were not welcome. Boyle called the Oxford circle "a knot of such ingenious and free philosophers, who I can assure you do not only admit and entertain Real Learning, but cherish and improve it." The Oxford circle had heard much about this extraordinary young man as well. Wilkins met Boyle in the summer of 1653 while Boyle was back in England for a few months. Wilkins lobbied him in person and then in a series of flattering letters, declaring that Boyle would "be a means to quicken and direct our enquiries." Finally, in 1655 Wilkins succeeded, and Boyle made his way to Oxford.

Boyle did not slip quietly into the Oxford circle but plowed into its midst like a man-of-war. He set up private laboratories and staffed them with mechanics, glassblowers, and apothecaries, along with personal secretaries who scribbled down millions of his words. He took over hosting the Oxford circle in his rooms, where he proved to be brimming with observations of his own, of other alchemists, of physicians, of iron-workers. Sometimes, however, it was difficult for Willis and the rest of the Oxford circle to understand what was on Boyle's mind. He swore his laboratory assistants to secrecy, and he kept some of his work cloaked in an alchemist's code, writing "copper" when he meant "gold," salting his notes with nonsense words, such as "Durca being freed from their sulphureous Baradam will afford Zahab or Kesph." Even when he was not speaking in code, it could take a while to extract his meaning. He was horribly disorganized, having claimed to have lost five hundred experiments at one go. Lorenzo Magalotti, an Italian who visited Oxford, found that Boyle "speaks French and Italian very well, but has some impediment in his speech, which is often interrupted by a sort of stammering, which seem as if he were constrained by an internal force to swallow his words again and with the words also his breath, so that he seems so near to bursting that it excites compassion in the hearer." He was no different speaking English. Boyle searched for precision and ended up with prolixity. And every time he mentioned the name of God, he paused for a moment and simply stopped talking altogether.

With patience, his new friends at Oxford could understand what Boyle was saying. It was well worth their effort. Boyle proposed to question the principles of chemical reactions just as the first generation of mechanical philosophers—including Descartes and Galileo—had questioned Aristotle's physics. At Oxford Boyle set out to demonstrate that Aristotle's four elements, his forms, and his qualities, could not account for them. He made his point with saltpeter, the mysterious stuff of both life and gunpowder.

Boyle showed that he could separate the components of saltpeter—two substances unlike saltpeter itself—and then recombine them. He first put a few ounces of purified saltpeter in a crucible with a flame underneath until it melted into a "limpid Liquor." Boyle added a burning coal, which made the saltpeter "boil and hiss, and flash for a pretty while." He added more coals until he could drive off nothing more from the crucible. Boyle called what was left when the crucible had cooled down a "fixed salt." It had none of the qualities of saltpeter. It was no longer white but blue-green. It had a taste and an odor, neither of which had existed before.

Mixing the fixed salt in a vial with water, Boyle then added spirit of saltpeter (known today as nitric acid). Alchemists created spirit of saltpeter by distilling saltpeter in an alembic, much as pure alcohol can be distilled from wine. His friends watched agog as the fixed salt and the spirit of saltpeter began to hiss, turned hot, and bubbled over. "Ice-icles" formed in the vial in a matter of seconds, and they proved to be saltpeter, reunited into its former self.

Boyle argued that he had just pulled apart and reunited the fundamental particles of matter. He did not use the word "atom," which Gassendi had used to refer to an indivisible unit of matter, preferring to call his particles corpuscles. By this he meant particles that resisted all human attempts to divide them—leaving open the possibility that someone someday might figure out how to split them into smaller particles still. Boyle declared that corpuscles had only a few properties—shape, size, and perhaps motion. Anything beyond that short list—colors, smells, tastes, sounds—was produced by the patterns created by groups of corpuscles as they linked to each other with hooks and eyes, a pattern Boyle called texture. Chemical reactions, like Boyle's saltpeter phoenix, occurred when different collections of corpuscles came into contact, liberating or trapping one another and creating new kinds of textures. By burning saltpeter with the coal, Boyle had pulled apart their constituent corpuscles and linked them in new combinations. When he mixed the fixed salt with spirit of saltpeter, the corpuscles joined together again into their original combination. Aristotle, Boyle argued, could not account for the return of the saltpeter in terms of his four elements, his forms, and his qualities. They were not explanations, only labels.

Alchemists, Boyle believed, had gone a long way toward figuring out the constituents of matter, but he didn't like the way they tried to limit them to just a few elements or principles, such as the spirit, sulfur, and mercury of Paracelsus. Those principles might turn out to be a mixture of corpuscles that alchemists did not yet know how to separate. Van Helmont, after all, had shown how many substances produced their own gases. There might turn out to be any number of different kinds of corpuscles. Boyle chose to call these kinds "elements," but not in the sense that Aristotle used. He used the word to describe any "perfectly unmingled bodies" that could not be broken down into some other substance. There might be ten elements or a hundred or a thousand—Boyle did not actually identify any by name. He was less interested in promoting a rigid system of elements than in finding ways to split what seemed unsplittable.

For Boyle, alchemy and Aristotelian philosophy both fell short not only in substance but also in style. Alchemists such as van Helmont and Paracelsus had disparaged reason, claiming that their knowledge came as revelations directly from God, which of course made it impossible for anyone else to test whether they were right.

Aristotle's followers depended too much on reason alone, building elaborate arguments out of ancient authority. Boyle was wary of becoming wedded to any system that might fail to encompass the full truth of nature. Instead, he championed the working hypothesis. His explanations for his experiments were always open to revision as new evidence turned up. "I dare speak positively of very few things," he wrote. The working hypothesis was not just a scientific procedure for Boyle but a religious act. He would never dare to claim he understood more than the rough outlines of God's will.

The Oxford circle was exactly the sort of environment in which Boyle felt natural philosophy could be explored best—a group of trustworthy, elite witnesses who could vouch for the reality of the experiments carried out before their eyes. He turned an informal club into a public testing place for scientific ideas, even welcoming lords and ladies to add more respectability to his experiments. When Boyle wasn't performing experiments, he was writing them down. He would make the public at large his witnesses as well, even if most people could never hope to test his claims. Boyle's scrupulosity—his compulsion to leave nothing out in his dealings with God—became a relentless attention to scientific detail, to documenting even experiments that failed, since he could never be sure that he had missed something of importance.

Boyle rejected Aristotle in favor of mechanical philosophy in a way profoundly different from Descartes's and Hobbes's. Those two philosophers claimed that matter was completely passive, moving only because of inertia and collisions. To many English natural philosophers, Boyle included, something more was called for. Boyle believed that God had launched into motion the corpuscles of the world at Creation, and that ever since, they had continued and passed on that motion to other corpuscles they crashed into. "It more sets off the wisdom of God in the fabric of the universe," he wrote, "that He can make so vast a machine perform all those many things, which he designed it should, by the mere contrivance of brute matter managed by certain laws of local motion and upheld by his ordinary and general concourse."

Boyle knew that churchmen might not trust this sort of philosophy, which seemed to allow the universe to purr along smoothly without God's divine touch. But Boyle argued that it would be impossible to discover God's plan from pure reason alone, because human minds could not grasp it. Experiments could reveal some of the grammar of God's language, Boyle declared, and that evidence would go far in revealing the reality of God to even the most skeptical atheist.

On all levels—from the very practice of science to the particulars of chemistry, physics, and medicine—Boyle gave a coherent voice to the Oxford circle's aspirations. In the late 1650s, that voice was not heard by many people beyond the Oxford circle itself. To the rest of England, Boyle would have seemed just a peculiar aristocrat with no university degree, no license to practice medicine, no position in government, no published work to his name—a man with little more than a passion for alchemy, which was not unusual for someone of his class. But he had already altered science for good.

Thomas Willis immediately took to Boyle. Having traded letters for several years already, they became friends and scientific partners, talking over their experiments and reading each other's unpublished essays. Willis sent his assistant Robert Hooke to work with Boyle, recognizing a good scientific match in the pair. Willis's instincts were right. Boyle had an army of assistants working for him, but Hooke quickly became Boyle's equal partner. It's said that Hooke even taught Boyle Euclid and introduced him to Descartes.

Boyle shared Willis's passion for medicine and anatomy. Obsessed as ever with his own health, or lack of it, he spent much of his time in Oxford experimenting with remedies of his own. He hoped to find the nature of the link between the weather and outbreaks of disease, setting Christopher Wren and Robert Hooke to invent weather clocks that traced a record of the winds and temperature on rolls of paper, in order to trace epidemics back to exhalations rising from the Earth.

Boyle and Willis also shared the conviction that in order to build a new kind of medicine, they would have to figure out how the body worked. Like Willis, Boyle had become a dedicated follower of William Harvey (on a visit to Harvey for his bad eyes, they had talked about his anatomical discoveries). "I have not been so nice," Boyle wrote, "as to decline dissecting dogs, wolves, fishes, and even rats and mice, with my own hands." He killed white rabbits so that he could hold their transparent eyes up to a window and see the panes and bars thrown in miniature onto their retinas. Boyle saw the body much as Descartes had—he liked to call it a "curious engine"—but he did not see an earthen machine of the sort Descartes imagined. "I look not on a human body, as on a watch or a hand-mill," Boyle wrote, "i.e., as a machine made up only of solid, or at least consistent parts; but as an hydraulo-pneumatical engine, that consists not only of solid and stable parts, but of fluids, and those in organical motion." A natural philosopher could not treat the body as a set of pulleys and levers. It was wet and mutable, a chemical factory that sucked in air and food and water and recombined their corpuscles like an alchemist in his laboratory, creating new forms. Only through alchemy, he believed, could its workings be known.

Boyle also hoped to find in corpuscles an explanation of life and how it differed from inanimate matter. Like Willis, he was distilling blood to break it down to its ultimate components. For Aristotle, the soul was the difference between life and non-life, but Boyle had abandoned those old intrinsic forms. What Aristotle had called a soul Boyle saw as a phenomenon that emerged from the organization of countless corpuscles. Death did not remove the vegetative or sensitive soul; it altered the texture of the corpuscles that made up the body. Boyle was impressed by the way that a dead man's nails could keep growing for months, long after the rational soul had left his body. Life and death for Boyle were no longer opposites but part of a single continuum of chemistry.

Boyle and Willis both were trying to redefine the soul through their work. Boyle objected to the way Aristotle and his seventeenth-century followers endowed matter with the passions and even intelligence of a soul. To Aristotle, sucking the air out of a reed dipped in water created a horror of a vacuum in the water, which responded by rising up to take the air's place. To Boyle this meant that the water "knows both that air has been sucked out of the reed and that, unless it succeed the attracted air, there will follow a vacuum; and that this water is withal so generous, as by ascending, to act contrary to its particular inclination for the general good of the universe, like a noble patriot that sacrifices his private interests to the public ones of his country." It was absurd to give water so much wisdom. Boyle found Plato's world-soul equally dangerous. "This hypothesis is near of kin to heathenism," he declared. Boyle argued that Nature seems to act with a purpose not because it has a thinking soul, but because God's thinking is at work in it.

Boyle did not mean to fence off human passions from the natural world. He was all too familiar with the way even "the bare remembrances of a loathsome Potion" could stir up his body, creating a convulsive feeling of horror in the pit of his stomach. The sight of a whirlpool could make a man stagger with vertigo; a scandalous comment could stir up the blood into a bashful blush; a sad memory could bring tears—or as Boyle put it in his clinical way, "a considerable quantity of Briny Liquor is excluded at the eyes." The human body was like a musket, an engine designed to be sensitive to delicate triggers.

But thought, Boyle declared, belonged only to God, angels, and rational souls, all of which were immaterial. "It must be a strange aggregate of atoms, that could be able to devise and demonstrate all the propositions of the tenth book of Euclid," Boyle wrote. He came to believe in the same division of mind and matter as Descartes, but he came to it without Descartes's arrogant certainty. Boyle was always modifying, qualifying, and wriggling away from dogma. Anyone who could explain to Boyle how matter could reason "will, I confess, not only instruct me, but surprise me too."

Chapter Seven

The brain of a boy who was "a fool from birth," in Willis's words. Willis dissected the brain after the boy's death to discover the physical sources of intelligence. This illustration appeared in *The Anatomy of the Brain and Nerves*.

Spirits of Blood, Spirits of Air

*I*t's no coincidence that shortly after Boyle's arrival Willis produced his first important scientific work. Boyle introduced Willis to his seminal ideas about experiments, alchemy, and corpuscles. In Willis's search for an explanation of fevers—and for the heat produced by all humans and animals—he had already abandoned Galen's ideas about humors and innate heat. Now he found a compelling explanation in chemistry. "I thought best, the common acception of humors being laid aside," he wrote, "to bring into use these celebrated principles of the chemists."

For Willis, the body had become a collection of atoms. "All Natural effects," he wrote, "depend upon the Conflux of Atoms diversely figured, so that in all Bodies there be Particles Round, Shape, Four square, Cylindrical, Chequer'd or Streaked, or of some other Figure." But Willis, like Boyle, did not claim to know the precise identity of these atoms. They were too remote from human perception to be useful to a doctor. Willis did know that when he put blood or urine over a flame, he could break it down into a few distinct substances. Whereas Harvey had treated the blood as a single spiritual substance, Willis now recognized that it was actually a chemical mixture. To the blood's components he gave familiar labels, some of which had been passed down from Paracelsus: earth, water, salt, sulfur, and spirit.

Like other English alchemists, Willis did not picture the particles that made up the body as Descartes did, like passive tennis balls bouncing and careening. For him, each principle had its own special types of motions and powers to make other particles move. Earth and water were the most passive, acting like a matrix for the other principles. Salt was more active but stable enough to give life weight and duration and produce new life. (Willis was struck by the fact that seeds were salty.) Sulfur and spirit, agitated particles in continual motion, were the active principles of life. Sulfur was the source of fire, Willis claimed, and spirit, the most active matter of all, consisted of "Aetherial Particles of a more Divine Breathing." Mixing even a touch of spirits or sulfur into the other principles triggered a life-giving transformation: a ferment.

Van Helmont had written about ferments just a few years before, but he saw them as citizens of the spiritual cosmos, divine souls penetrating matter and transmuting it into new forms. Willis, like Boyle, chose to strip the mystical from van Helmont. He wanted to explain the body "without recourse to occult qualities, sympathy, or other refuges of ignorance." Ferments for him were not something that happened out on the astral plane. They were a dance of particles. They were the everyday business of farms like the one he had grown up on. He knew that leaven could transform bread dough from a sticky, moist lump into an airy loaf. Willis declared that particles of leaven raced around the inside of the dough, lifting it up and creating holes as they stirred about.

Blood was a ferment as well. If Willis let it sit in a bowl, it separated into different parts—a bright red liquid at the top and a purple, almost sludgy fluid below. In time, a clear, watery substance would rise up, which, if Willis put a flame under it, would turn into white solids.

The ferment of the blood produced heat just as yeast did in dough. Willis declared that the heating took place in the heart, where a ferment loosened the bonds that held the particles of sulfur, salt, earth, and spirit together in corpuscles. The liberated fiery sulfur made the blood "leap forth with a frothy heat" when it entered the arteries. Unlike dough or beer, which had ferments that sooner or later came to an end, blood was in perpetual ferment, as new blood was created and old, corrupted blood was destroyed.

In corpuscles, Willis found a new, elegant way to explain fevers. Van Helmont had claimed that fevers arose when some kind of unwanted matter entered the body and the archeus tried to shake it off, inflaming itself in the process. Willis replaced van Helmont's spiritual struggle with a mechanical story. Fevers, he said, were caused when something disturbed the regular fermentation of the blood and stirred up its particles violently. The blood then started to boil like a pot on a fire, pushing against the blood vessels, creating a faster pulse, and sending out burning heat through the whole body.

Medicine itself, Willis decided, was the study of human fermentation. "We are not only born and nourished by the means of ferments, but we also die," he wrote. "Every disease acts its tragedies by the strength of some ferment." A doctor should manage his patients in the way a vintner manages his aging wine bottles, checking their chemistry and altering it to guide its fermentation. "For to the preserving or recovering the health of man, the business of a physician and a vintner is almost the same," Willis said.

In his synthesis of Harvey's anatomy, van Helmont's alchemy, and Boyle's corpuscles, Willis helped to invent something that would ultimately become biochemistry. Today, scientists know that the body's heat is produced by a very specific sort of fermentation, as enzymes chop up food, build proteins, copy genes, and take out the metabolic trash from our cells. These chemical reactions release heat, which keeps the body's temperature at a steady level. Diseases cause fevers by altering that same balance of chemical reactions, so that overall they produce more heat. Willis took a crucial step from Harvey's work toward this understanding. And in the process he helped dismantle one part of the traditional triad of souls: the vital soul of the heart that was supposed to supply an innate heat to the body.

By 1656, Willis had worked out his theory of fevers far enough to begin writing a book, which he circulated around the Oxford circle. Boyle worried that Willis was putting too much stock in the notion of spirit and sulfur and the other principles without really knowing what they were. "This, I fear, may prove somewhat prejudicial to the advancement of solid philosophy," he later wrote, but nevertheless he passed Willis's manuscript on to his own network of friends. Well before it was published in 1659, it stoked Willis's reputation, earning him praise as "a man uncommonly skilled in philosophy and Hermetic medicine."

As Willis's book on fevers crept slowly toward publication, his own patients' fevers kept him busy. New epidemics of influenza, smallpox, and meningitis swept through Oxford, driving poor and rich alike to Willis. Willis designed his own drugs, such as a mysterious steel syrup, and he had his apothecary live with him in Beam Hall to keep his remedies secret. He had come a long way from swirling vials of urine in village markets.

Willis's business became so good, in fact, that he felt ready to marry in 1657. He chose Mary Fell, in his choice showing just how deeply his royalism dyed his life. Mary was the sister of John Fell, a fellow defender of Oxford in the siege and now a priest in the Loyal Assembly. She was the daughter of Samuel Fell, who had been dean of Christ Church before the war.

No picture of Mary Fell survives, nor do many words, but what little does exist reveals a woman who inherited her family's tough character. Nine years earlier, her father was thrown into jail by Parliament, and the Visitors from Parliament told his family to leave their rooms at Christ Church so that they could install a handpicked dean of their own. Mary, her brothers, and her mother refused, locking the doors on the Visitors and their soldiers. A crowd formed to watch the confrontation, and it almost certainly included Thomas Willis, who lived nearby at the time. The Visitors had the doors of the Fells' rooms broken in and told the family to leave. They still refused to move. A guard was then placed over them with instructions "to weary them out with noise, rudeness, and smell of tobacco," according to the diarist Anthony Wood. Eight days of this psychological warfare proved useless, and so the new chancellor ordered guards to take out Mary and her family physically. "The Chancellor desired Mrs. Fell to quit her Quarters, telling her that in so doing she would do God and her Country good service," Wood wrote, "but she refusing that kind of proposal, had very ill language first given to her by him, and then she was carried into the Quadrangle in a chair by soldiers. Her children were carried out upon boards." After this humiliating eviction, Mary came to Willis's house to pray with the Loyal Assembly, and after nine years, Willis finally asked her to marry him.

Willis became prosperous enough to support a growing family of children and a crew of servants. His star rose even further when his fever book finally came out in 1659. It was popular not only in medical circles in England but across much of Europe as well. He became a controversial, celebrated doctor striking a powerful blow against Aristotle's four elements and Galen's four humors by demonstrating that corpuscles could explain the most important threat to human life. Traditional physicians were appalled that anyone would dare barge into the supreme philosophy of medicine with the sooty products of an alchemist's laboratory, to claim that the spirits of the body and the spirits distilled in an alembic were the same. They attacked "the lechery of the wanton mind and the mad itch for innovation" in Willis's book, which they were convinced would "end in the ruin of the human race."

But others praised Willis for his challenge to conventional wisdom. "We should rather reject what runs to superstition, and not pin the Faith of mankind upon the sleeve of Hippocrates because others have done so," one of them declared, calling his book "an excellent example of Bacon's idea of raising a philosophy upon sensible experiments." For the next century, physicians and apothecaries would subscribe to Willis's theory of fevers almost word for word.

Paradoxically, one reason that they held onto it for so long was that Willis explained the cause of fevers in a new way but he didn't change their treatment. "The opening of a vein," he wrote, "cools and ventilates the blood, as by that means, 'tis less torried, or scorched, and is circulated more freely in the vessels, without danger of burning." Willis was gentler with his bleedings than traditional physicians, prescribing them only in the early stages of a disease. He added Paracelsist, metal-based drugs to his pharmacopoeia, but used them to do the same things that Galen had done—to purge, heat, and void the body. Ultimately, Willis proved to be a backward-looking revolutionary: he preserved the old medicine within the new science.

As Willis cared for the victims of the new fevers in the late 1650s, he was forced yet again to see how the mind could be altered by disease. People stricken with meningitis "talked idly, and at random," he wrote. Sometimes, as they declined toward death, he watched his patients lie "for the most part without speaking, or knowing those about them, as it were stupid, and it rarely happened in this fever, that anyone about to die, was so perfect in their memory and intellect, as to dispose of their family affairs, or to take leave of their friends."

For Willis, nothing could be quite as frightening as the loss of one's intellect. He rejected the Puritan creed that everyone's fate was predestined, believing instead that salvation came only to those who lived well—with repentance, obedience, and duty. Those who lacked their reason couldn't live out this sort of existence and might not get into heaven. The priests who were still preaching secretly in Willis's house declared that medicine was valuable because it could prolong the life of mortally ill people long enough to convert and find eternal life, but if diseases robbed his patients of their rational minds, they would be lost.

Worrying over the souls of his patients, Willis began to think more seriously about the animal spirits. Convinced that spirit was the most active principle of nature, he believed animal spirits were the most active of all spirits. If life depended on fermentation, then animal spirits would be the most vital substance of all in the body, which would make the brain and nerves the most important organs of all. In Willis's book on fermentations and fevers, he touched on animal spirits, writing that their "origins and motions are very much in the dark. Neither doth it plainly appear, as to the animal spirit, by what workman it is prepared, nor by what channels it is carried, at a distance, quicker than the twinkling of an eye."

The alchemist in Willis began to speculate. The brain, the skull that capped it, and the nerves that descended from it all looked to him like "a glassy alembic, with a sponge laid upon it, as we use to do for the highly rectifying of the spirit of wine." Willis imagined that when the blood rose to the brain, the spirits were distilled out of it, leaving behind the heavier particles. The brain then soaked up the spirits like a sponge. Rather than being injected into the ventricles, they entered the brain itself. Once liberated from the blood, animal spirits were prone to escape away as vapors, and so they could only flow through tiny channels—invisible pathways through the brain and then the nerves, which sprouted from the brain like the necks of an alembic.

Crude as this image may be, it represented Willis's first step toward the modern science of the brain. He grew more and more curious about this marvelous alembic, and now he had the unusual luxury of studying the brains of his dead patients. Even a few years earlier, most of the bodies that anatomists dissected belonged to criminals. As Willis became more famous and anatomy became more respected, however, he was able to persuade families to let him cut open his dead patients. Willis wanted to link the diseases that had killed them to unusual features of their corpses. In some of the victims of the meningitis outbreak, for example, Willis discovered that the brains were coated with a thick, bloody mass. In cases of other illnesses, however, he did not find the links he had expected. Deliriums and frenzies were thought to be caused by a damaged diaphragm, that barrier to the lower soul, but during an autopsy, Willis discovered a hole in the diaphragm of a man whose mind had been sound his entire life. Willis wondered if the traditional explanation for delirium and frenzy was mistaken. In his autopsies, Willis also saw something else: the conventional pictures of the brain were wrong. Even Vesalius had made a mess of the brain. The problem, Willis suspected, had to do with the way anatomists normally examined it, by slicing off one layer after another as the brain rotted away.

The only way to rectify these mistakes, to get at the secret nature of the animal spirits, would be to make a proper study of the brain and the nerves. It would take Willis years to launch this project. His delay was not a matter of intellectual apathy or lack of courage, as he once put it, to "wander without a leader or companion in many out of the way places, and trample as it were in a solitude trodden by no footsteps." It's possible that Willis didn't investigate the brain further because he would inevitably have been led into an investigation of the soul—a truly dangerous pastime for a man like Willis, with a house full of royalist conspirators, a despised theology, and no university post. The late 1650s would have been a particularly bad time for Willis to start this investigation, because the fever that had gripped England for almost twenty years was spiking once more, with a ferment of agitating particles stirring its blood.

Oliver Cromwell had become an old, ill man. Once he made himself Lord Protector, the fire of revolution seemed to die within him, and he simply wanted to do whatever he could to enforce peace. When he tried to create a spirit of religious toleration in England, however, he found that his subjects had other ideas. Anabaptists were stirring up Oxford so much that Cromwell had to send soldiers to the city to bring them under control. After a leading Quaker rode Christ-like into the city of Bristol on a donkey with women casting palms before him, Parliament had his tongue bored through and his forehead branded with a *B* for "blasphemer," then had him whipped as he was paraded around London.

The year 1658 arrived with evil portents: a splintering winter, an outbreak of fevers. That August, Cromwell was devastated when cancer dragged his daughter Elizabeth to death. Shortly after her funeral, George Fox, the founder of Quakerism, came to see Cromwell to speak on behalf of his suffering people. He met the Lord Protector riding at Hampton Court. "Before I came to him," Fox later remembered, "I saw and felt a waft of death go forth against him, and when I came to him he looked like a dead man." Cromwell soon came down with pneumonia and died on a night filled with storms that brought down steeples and uprooted trees.

For all the turmoil of his reign, Cromwell had done much for his country. England had become more powerful than ever, with growing colonies in the New World and respect from the great nations of Europe. But he had failed to build a government that would survive his death. In his place, his son Richard was thrust. Oliver had kept Richard safely tucked away in the countryside for years, so that he never gained a shred of military or political experience. Now thirty-two years old, Richard Cromwell came to London to rule over a Parliament that was millions of pounds in debt and a country now seething with competing factions. Political radicals found their strength again and demanded that the army hand out unfarmed lands to the poor. Religious sects swarmed through London's streets, using the chaos as a chance to save England's soul. "Oh priests, howl and weep, for a day of lamentation is coming upon you, a day of woe and misery if you repent not," warned the Quaker William Simpson.

The fanaticism and confidence of the mob frightened the conservative landowners. One lord complained that the goal of the Quakers and Anabaptists "is only to turn out the landlords." Powerful men began to grumble the unthinkable: England could be tamed only if Prince Charles returned from exile to take on his father's throne. The leaders of the army were determined to keep the country a republic, however, and were ready to crush any royalist uprising.

Anxiety rose throughout the country, and the people of Oxford became especially tense. When the mayor announced that

Richard Cromwell had become Protector of England, he was pelted with carrots and turnip tops. New rules were put in place to keep the university godly: no spurs, no hair powder, no tennis. The authorities established daily marathons of sermons, which royalists skipped to go to coffeehouses. Orders for an uprising were coming from abroad and Cromwell's soldiers went from house to house in Oxford searching for arms.

Fanatics were also rising against Oxford again. George Fox attacked the university in 1659. "This is a call for every uncircumcised Philistine"—in other words, Oxford scholars—"to come forth into the open field and there to be tried with little Davids who have the bags and the slings, and the stones, which is the power of God." In a windstorm in July 1659, some stones were blown from the top of St. Martin's Tower in Oxford. At just that moment, a trumpet was sounding to summon parliamentary soldiers for another search for arms. Inside the church, the sound terrified people in the gallery. Suddenly convinced that the Quakers and Anabaptists were coming to slit their throats, they cried murder and leapt down in panic on the congregation below.

The times were certainly not good for radical ideas about the brain and the soul. But Willis's hesitation may not have flowed entirely from political fears; he also lacked some crucial tools for the job. The brain, after all, was the most challenging organ for an anatomist to study—hard to reach, easily damaged, and quick to decay. For the time being, Willis and his friends remained content to investigate the blood. And in that investigation they developed much of the technology Willis would use a few years later to begin his study of the brain—preservatives, microscopes, and injections.

—

During these uncertain times, Willis ushered another great mind into the Oxford circle. Richard Lower, the son of a Cornish gentleman, came to the university in 1649, and his medical gifts came to Willis's attention a few years later. Although Lower would not get an official medical degree for another ten years, he was soon acting as junior partner to Willis, dispatched to see patients as far away as Cambridgeshire. When Lower wasn't treating patients, he was usually dissecting some creature. On a Sunday morning, he might pass up church to cut open a calf's head in his rooms. Lower stood in Willis's shadow for a decade, taking fees that were a fraction of his master's. He discovered a spring near the village of Astrop that spurted water with healing powers, and in later years the sick would flock there to drink it, but Willis got the credit for its discovery. Lower wasn't bitter, though. Willis trained him well, guided him to a wealthy medical practice, and spoke highly of him whenever he had the chance. In his books, Willis declared his gratitude for the edge of both Lower's knife and his wit.

Early in their collaboration, Lower helped Willis with a question that had emerged from his work on fevers: what determined the color of blood? Willis knew that veins carried dark blood, while the blood in arteries was bright red. Galen had claimed that the veins and arteries were separate systems of vessels with distinct supplies of blood, but Harvey had shown that Galen was wrong. Willis's own research had led him to decide that the heart contained a fermenting fluid that transformed blood as if it were wine or beer: sulfur, salt, and spirit mixed together in the blood and turned it red. The experiments Willis and Lower carried out seemed to confirm his idea. They poured blood from a vein, dull and purple, into a shallow bowl. Before long a bright red layer formed on the surface of the blood. That was exactly where Willis expected the blood's light, violent, sulfurous corpuscles to end up.

But Boyle and Hooke were at that moment discovering the first hints that Willis and Lower had missed the mark. Over the course of a decade, those clues would lead the Oxford circle to discover that the blood's power came not from the heart but from the air itself. They were not reviving Plato's world-soul, though. They had discovered instead a new sort of corpuscle mixed invisibly into the atmosphere.

Boyle had been looking for evidence to destroy Aristotle's claim that nature abhorred a vacuum—as if it had a soul that acted on its own desires—and he found it in the barometer. The barometer at the time was a new invention, created by a disciple of Galileo's named Evangelista Torricelli. Torricelli had been intrigued by the fact that no suction pump could lift water higher than thirty-two feet above the ground. If the water was rising inside the pump thanks to nature's horror of a vacuum, why did that horror always stop at the same height? Torricelli declared that the horror was an illusion. Instead, air was a substance with bulk. It might be far more delicate than water or stone, but its weight pressed down on the Earth, and also the water in a pump. Sucking the air out of a pump removed an invisible cap on the water inside it. The force of the air pressing down on the surrounding water would then push up a rising column of water inside the pump until the weight of the column balanced the weight of the downward-pushing air.

If that were true, Torricelli argued, a denser liquid should stop rising at a much lower height. In 1647 he filled a tube sealed at one end with mercury, which is fourteen times denser than water. He tipped the sealed end up and held the tube upright in a dish. Some of the mercury flowed down into the dish, but the rest remained behind as a column in the tube. Just as Torricelli had predicted, the column's height was a fourteenth the height of a column of water. "We live at the bottom of an ocean of the element air," Torricelli declared.

Mercury-filled tubes—otherwise known as barometers—became a sensation throughout Europe. Mountain climbers took them on their hikes and watched the columns shrink as the air around them thinned. The Oxford circle was fascinated by them as well. Christopher Wren realized a barometer could test Descartes's claim that the universe was filled with whirlpools of particles. Descartes had argued that the moon caused ocean tides by pushing against the particles between it and the Earth. If that were true, Wren observed, a barometer should fluctuate in sync with the moon's orbit. Boyle carried out the test by setting up a barometer and leaving it in place for weeks. It slowly rose and fell two inches, but not in the moon's rhythm. No one knew what was making the air heavy and then light, but it was certainly not the moon.

In 1658, Boyle heard of an even better instrument for studying the weight of air. A German burgomaster named Otto Guericke had built pumps that could suck air out of closed containers. In his most spectacular experiment, Guericke built two bowls of copper that he clamped together into a fourteen-inch sphere. He then attached a suction pump to it and slowly drew out the air. The bowls became sealed so tightly in the process that two teams of horses couldn't pull them apart.

When Boyle heard about Guericke's pump, he instantly wanted one of his own, but one that would be better suited to his style of experimenting for his friends. Guericke's pump had to be worked by two strong men for hours, and its solid metal walls hid what was happening inside. Boyle asked Hooke to apply his mechanical genius to the problem, and Hooke did not let him down, building a transparent globe of thick glass mounted on a wooden frame. He added a crank handle to draw the air out of the globe quickly and a lid-covered hole through which he could put objects inside.

When Boyle laid eyes on Hooke's creation, he immediately wrote out a list of experiments to run, a list so long that it look him years to finish them all. Boyle and Hooke brought their air pump before the Oxford circle and put on a performance that no one had seen before. When Boyle placed a barometer inside it, they could see the column of mercury drop as the handle was cranked. When Boyle flattened a lamb's bladder and put it inside, it inflated as the air surrounding it was sucked out. Boyle's audience watched smoke flow like water and wine boil at room temperature. Candle flames suddenly sputtered and died. Boyle used the pump to calculate that air was about a thousand times lighter than water, meaning that the atmosphere must extend a hundred miles above the Earth—both of which would turn out to be not far from the truth.

The miles of air overhead, Boyle speculated, pressed down on the particles near the Earth, which were squeezed like springs in a clock.

By pumping air out of his globe, Boyle removed that weight, and the particles were liberated, pushing out against the walls of a bladder or escaping from wine. Boyle did not believe that his pump experiments let him choose between the two most popular theories about corpuscles—Descartes's whirlpools of particles packed together cheek to jowl or Gassendi's lonely atoms gliding through the void. But it did clearly show one thing: the horror of a vacuum was no horror at all but simply the spring of air pushing against everything on Earth. Hooke even suggested that the air was an invisible network of springy particles like an invisible wool. Boyle, as ever, hesitated to settle on any one explanation for what he saw.

Thomas Willis was among the privileged few who watched these magical feats. As extraordinary as the experiments were, he must have been astonished most of all when Boyle and Hooke placed a lark in the globe. As the air was pumped away, the bird began to appear sick. It went into convulsions. Within ten minutes it was dead. The same thing happened when Boyle and Hooke experimented with a hen-sparrow and a mouse, with snakes and shrews. The vacuum pump could withdraw some basic requirements for life.

Everyone knew that all animals must breathe to live. Galen had offered a reason why, declaring that lungs attract air, which then passes into the heart, creating the vital spirits of the arteries and the animal spirits of the brain. At the same time, the air also cooled the hot heart, and with each exhalation, the lungs released the blood's sooty fumes that would otherwise become poisonous. No one challenged Galen's ideas with any real alternative until 1654, when Ralph Bathurst, Willis's partner in medical poverty, argued that lungs did not move of their own power. When the diaphragm moved downward, it somehow drew air into the lungs.

Bathurst pointed out that air was not essential for cooling the heart and blood—fish, after all, had no lungs at all, and their blood and hearts were cold. Instead he proposed that the same particles of saltpeter that made gunpowder explode were actually suffused throughout the universe, a chemical version of Plato's world-soul. When an animal breathed in air (or drew water through gills), the particles were distilled in the body's internal laboratory, where they fueled a living fire.

Boyle had listened to Bathurst's ideas when he first came to Oxford, and now he found support for them with his pump. As the lungs expanded, it now became clear, the pressure of the air inside them became lower than outside. As a result, air automatically rushed into the lungs. When Boyle pumped the air out of his chamber, there were fewer particles left to move into a lark's lungs. "This invited us thankfully to reflect upon the wise goodness of the Creator," Boyle wrote, "who by giving the Air spring, has made it so very difficult, as men find it, to exclude a thing so necessary to Animals."

Boyle agreed with Bathurst that the lungs did not serve to cool the blood. Nor could the lungs' sole purpose be to get rid of fumes. If that had been the case, the lark would have thrived in the glass globe. Boyle was struck by the fact that sometimes animals deprived of air would collapse but not die. Even after a few minutes, he could pump the air back into the globe and watch a bird revive. Without air, animals were like machines deprived of their power, he wrote, "such as is offr'd to a Wind-Mill when the Wind Ceasing to blow on the sails, all the several parts remain moveless and useless, till a new Breath put them into motion again."

"There is some use of the Air," Boyle decided, "which we do not yet so well understand." Perhaps a lark and a flame both needed the same substance in the air, which was why they both died in Boyle's pump. Perhaps there were particles mixed among the air that were essential to the vital spirits.

History has had a difficult time taking in the full scope of the works of people like Robert Hooke, who is remembered as an inventor and a physicist, or Robert Boyle, the first modern chemist (or the last alchemist, who brought alchemy's reign to a close). It's as if there isn't room in our historical memory for the fact that they also began to crack the mystery of breathing. In the Oxford circle, however, there were few boundaries. Christopher Wren, for example, is remembered today as a great architect, but in the late 1650s the Oxford circle knew him as an astronomer who could take the spleen out of a dog.

By his early twenties, Wren was already famous for drawing exquisitely detailed maps of the moon, as intricate as any map of Earth, which gave Galileo's squinty visions a reality full of mountains and valleys. He built giant telescopes, and discovered "arms" around Saturn that would ultimately resolve themselves into rings. In his mid-twenties he spent two years as professor of astronomy at Gresham College in London before taking up the same post back at Oxford. But during that time, Wren also applied his genius and his agile fingers to the workings of the body. He dissected the eyes of horses to follow the path of light as it was refracted through lens, iris, and crystalline humor. He then built the first scale model of an eye.

Looking through a microscope, Wren drew titanic fleas and ants. He looked at the microscopic details of nerves and claimed to see "little Veins & Arteries in them." Sadly, none of Wren's original drawings from these projects has survived, although judging from his other medical artwork that has, they must have been sublime. He spent a lot of his time in Beam Hall during his Oxford years, watching Willis perform dissections and painting delicate watercolors of the strange things that emerged. Wren could even find beauty in an ulcerated intestine.

Wren also carried out surgical experiments. One of the many questions that emerged in the wake of Harvey's work was the function of the spleen. In the ancient four-humor theory, melancholy (otherwise known as black bile) was thought to be produced in the liver, whereupon it was attracted into the spleen and from there voided into the stomach. Too much black bile caused diseases, depressions, and delirium. Without a spleen, the black bile ought to pile up in the body to dangerous, perhaps lethal levels. But in the early 1650s, some of Harvey's followers looked closely at the blood vessels attached to the spleen and discovered that none of them linked it to the liver or stomach. Exactly what the spleen did they couldn't say, but it couldn't involve melancholy.

Willis speculated the spleen had something to do with the fermentation of blood. "It seems that it is, as it were, a store-house for the receiving of the earthy and muddy part of the blood, which afterward, being exalted into the Nature of a ferment, is carried back to the blood for the heating of it," he wrote.

Wren decided to see what would happen if he removed a spleen from a dog and closed off the severed arteries. Boyle gave Wren one of his dogs to operate on, which he lashed to a table. He placed a cushion underneath its belly, clipped away a patch of fur, and drew a line two fingers below the dog's rib cage at right angles to the muscles of the abdomen, to mark where he would make his incision. Taking up the sort of knife farmers used to castrate their pigs, he slipped the blade into the dog. But Wren pierced only the muscles and the lining of the abdominal cavity, taking care not to nick the dog's intestines. He ripped along the line he had drawn to its end. Pressing down on the abdomen with one hand, he slipped the dark slab of the spleen out of the opening, and then tied off each of the veins and arteries that supplied the organ. With the spleen now removed, he dabbed the tied ends of the blood vessels with balsam. Wren sewed up the wound, and then he covered it with a plaster and swathed the dog to keep it warm while it recuperated. He left a small opening in the dog's belly so that he could check for clotted blood and regularly wash out the wound with a decoction of barley, honey, roses, and red sugar. Although he had no antibiotics or the other tools of modern surgery, Wren performed a flawless operation. Two weeks later, the dog was back to normal, even without its spleen. Boyle reported with satisfaction that it was "as sportive and as wanton as before."

But Wren's splenectomy was, in the end, little more than a demonstration of his steady hand. He did not bother to record his results or use it to build a new theory of medicine, let alone figure out what the spleen was really for. Wren liked to conduct spectacular scientific research to pass the time, the way some people make paper airplanes. We know of his splenectomy only through the marveling accounts of his friends. But by the time they were telling the story, Wren was already thinking about his next operation: to inject poison into the bloodstream of a dog.

The notion of making injections into the blood, like that of removing a spleen, had rarely been imagined before. For doctors who followed Galen, it seemed pointless. Yet injections would have a powerful effect if Harvey was right. And Harvey himself had carried out a crude precursor to injection, fastening a tube into the pulmonary artery of a cow and then using an ox bladder to inject water into it. The water flowed into the chambers of the heart, but none of the water managed to breach the chamber walls. Harvey never wrote about the experiment, but he did mention it to other physicians, who may have mentioned it in turn to Wren. In any case, Wren realized that if the blood circulated as Harvey claimed, an injection anywhere—even in a vein in the leg—would soon be carried throughout the body and its effects would be felt in all the organs.

As Boyle and Wren were talking one day about how poisons worked, Wren declared that he could deliver a poison by injection. Boyle was intrigued. In his days before Oxford, he had been perplexed by poisons. A snake's body was not poisonous, as Boyle learned when he fed a dog a chopped-up viper. The dog, he wrote, "liked his entertainment so well, that he would afterward, when he met me in the Street, leave those that kept him to fawn on and follow me." If blood was made from food, shouldn't the poisonous viper have killed the dog? Perhaps van Helmont was right, Boyle thought, when he had written that a snake's poison was actually its rage and fury—a spiritual venom. Boyle happily volunteered to help Wren carry out some experiments.

Wren had to invent the entire procedure for an injection from scratch. Boyle provided another large dog and rounded up Willis and some other doctors to help. Together they held the dog down and lashed its paws to a table. Wren cut open the dog's hind leg and made an incision in one of the veins. He tied it off and then produced a small plate of brass "almost of the shape & bigness of the nail of a man's thumb, but somewhat longer," Boyle later recalled. The plate had four holes at the corners, through which Wren passed a thread to attach it to the vein. It also had a slit running across its top, through which Wren cut an opening into the vein. Since the vein was tied upstream from the plate, blood didn't immediately come spurting out. The cut in the vein was big enough to insert a slender pipe, and through it Wren injected a mixture of opium and Canary wine. He then removed the brass plate, carefully stitched the vein, and closed the wound.

"When his vein was closed," Willis wrote later, "the Dog ran about as he used to do, seeming to be little or not at all affected with it: but after a quarter of an hour, he began to be a little dozed, to nod his head, and at last to fall asleep."

"There were wagers offering his Life could not be saved," Boyle wrote. A servant lashed the dog and forced it to run up and down in a nearby garden.

"At last by that means his sleepy inclination was quite off of him, and he became very sound and lively," Willis wrote.

"Having made him famous," Boyle complained, "he was soon after stolen away from me."

Wren continued perfecting his injections. He discovered how to hold veins in his fingers so that he no longer needed to use a plate to avoid tearing them. He switched from a syringe to a bladder attached to a quill. He began to inject water into dogs and then tried beer, milk, broths, wines, even blood. Some of his injections made the dogs stagger drunk or piss for hours; often they died. Sure that he had discovered something that "will give great light both to the Theory and Practice of Physick," Wren was not very bothered by the pain he was causing. To him, a few dogs were worth the sacrifice. In August 1657, while Wren was visiting the French embassy, the ambassador offered up "an inferior Domestick of his that deserv'd to have been hang'd." Wren tried to inject an emetic into him, but the servant fainted as soon as Wren brought the quill toward him—"either really or craftily," Boyle later complained.

Wren and Boyle were so happy with their results that they thought of other ways in which they could use injections. In the

1650s, cadavers spoiled, organs rotted, blood hardened. In death, the details of living bodies blurred away. Among his experiments, Boyle had tried to preserve organs and bodies in different liquids. He found that pure alcohol—what he called "spirit of wine"—worked best, "for this liquor being very limpid, and not greasy, leaves a clear prospect of the bodies immers'd in it," he wrote. Soaked in alcohol, bodies and organs held their original shape for years and could be sliced open to study. Boyle preserved linnets, worms, fish, caterpillars, and a human fetus. Now he experimented with injections as well, to see if "there may be some way to keep the Arteries & the Veins too, when they are empty'd of Blood, plump, and unapt to shrink over much, by filling them betimes with some such substance, as, though fluid enough when it is injected to run into the Branches of the vessels, will afterward quickly grow hard." With injections he wanted to freeze life.

In the final years of the 1650s, these injections brought more fame to the Oxford circle. Poets wrote of how they were "finding out the blood's Maeandring dances" and might even be able to inject blood from one person to another. The notion had existed since ancient Rome, when Ovid recounted the myth of Medusa giving youth to the old man Aeson by draining his blood vessels and filling them with a magical liquid. A few Renaissance physicians had contemplated transfusing blood, but without understanding how blood circulated, they could never make it a reality. Perhaps now Wren could.

Wren carried out his visionary work just as England itself was about to receive a fresh transfusion of royal blood. After Oliver Cromwell's death, his son quickly proved unable to master the anarchy of London, and in 1659, the army banished Richard to the countryside for an early retirement. Now the country was without any leader whatsoever. The factional chaos got worse, its echoes rolling north through England and over the hills of Scotland until they reached the ears of General George Monck. Monck was a practical-minded soldier who had served both Charles I and Cromwell, who put him in charge of the English troops in Scotland. Above all other things, Monck believed in order, and he realized that with Richard Cromwell gone, there was none left in London. He sent a letter to the army's general council condemning its attacks on Parliament and then led seven thousand of his men south to the capital to restore order. No faction was strong enough now to oppose him, and Monck arrived at Parliament's door without firing a shot.

Monck brought some short-term order to London, quelling the Quaker riots, throwing out the leaders of the army, and shutting down the pamphleteers who rallied the troops to radical causes. But the only way to bring England's balance back for good, he decided, was to restore its king. Monck invited back members of Parliament whom Cromwell had thrown out, including many who looked favorably on the monarchy.

After ten years of failed uprisings, Prince Charles recognized that Monck was offering him a chance to take his father's crown. Knowing that Protestants would be suspicious of his Catholic mother and hosts in France, he moved his court to Breda, a rock-solid Dutch Protestant city. He negotiated with English leaders rather than bullying them, promising amnesty to Parliament's veterans of the Civil War and giving Parliament a final decision on many matters, including religion. He would not make his father's mistakes.

His offer accepted, Charles entered London on March 29, 1660. His new subjects saw a thin thirty-year-old man before them, with glossy dark hair, a narrow moustache, and very tired eyes. He was accompanied by hundreds of gentlemen in ostrich feathers, velvet coats, and doublets made of silver cloth. Cannons blasted, guns fired, and bells rang, not just in London, but across the entire country. In Oxford, the city's fountains spouted claret and beer. Anthony Wood declared, "the world of England was perfectly mad."

But to an old royalist soldier like Thomas Willis, the opposite was true. It was as if England had finally gotten its brain back.

Chapter Eight

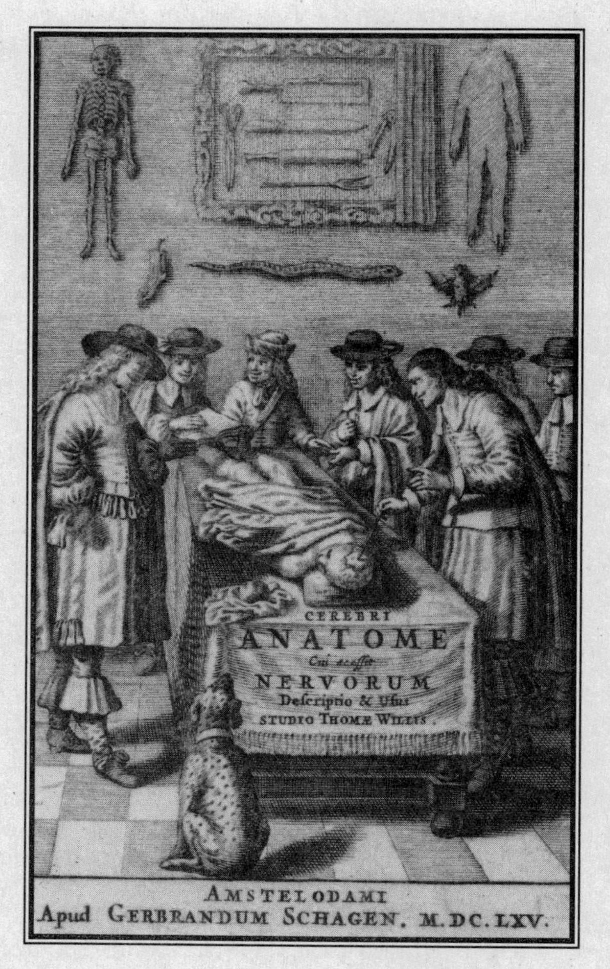

THE FRONTISPIECE TO THE 1665 LATIN EDITION OF *THE ANATOMY OF THE BRAIN AND NERVES.*

A Curious Quilted Ball

England got back more than just a king in 1660. The new Parliament elected that year was packed with old royalists determined to return English life to 1640. They made it illegal to gather a petition of more than twenty signatures and established fines for attending any unorthodox gathering. Squatters could now be thrown off unused land, while landowners could enclose forests and remove timber whenever they pleased. Many of the men who had signed Charles I's death warrant were tried for treason. Two of them were dragged through the streets with the severed head of another facing them the entire way. Oliver Cromwell had been dead for two years, and so his enemies did the next best thing to executing him: they dug up his embalmed body, let it hang for a day, and then stuck the head on a pole on top of Westminster Hall. It would look down on Parliament for twenty-five years, until the end of Charles II's reign in 1685.

Bishops and priests filed back into their old churches, carrying wounds that ran deeper than Parliament's. Archbishop Laud had been executed as a traitor, and many bishops had died in obscurity and poverty. After nearly twenty years of neglect, many cathedrals were in ruin. Some had been converted into stables, others into warehouses, taverns, and jails. Now that the Puritans had been overthrown, the Church of England would allow no compromise. Every English Protestant would take the sacrament, read from the Book of Common Prayer, stand for hymns, kneel for prayer, keep his hat off. The new bishop of London, Gilbert Sheldon, emerged as one of the church's toughest enforcers. He purged a third of the ministers in the city for refusing to submit to the new rules and shut down most of the printers who prospered under Cromwell. Every book and pamphlet published in England now had to win the approval of his censors. Anonymous pamphleteers were hunted down and thrown into jail. The exuberant debates that had flourished under Cromwell's reign shriveled away, leaving the public discourse of England bland and terrified. The only two literary masterpieces to come out of England during the 1660s, *Paradise Lost* and *Pilgrim's Progress,* were the work of a political outcast and a political prisoner.

The Puritan culture that had dominated England for fifteen years dwindled away, but even with the Puritans conquered, the bishops still felt threatened. Quakers and other so-called fanatic sects that had grown strong under Cromwell now defied Parliament's new raft of laws. Everywhere the Quakers seemed to be causing trouble—in the streets, where they refused to doff their hats, and in the churches, where they stood up during services with candles in hand, offering to burn parts of the Bible they found offensive.

As dangerous as Quakers were to the Church of England, at least they believed in God. Even more terrifying to the bishops was the prospect of Englishmen giving up altogether on their Lord and their own immortal soul. Priests warned their congregations again and again about the dangers of atheism. The notion that atheism was rampant in England became rampant itself—Anthony Wood bemoaned living in "an age given to brutish pleasure and atheism."

But historians who search for evidence of these atheists—pamphlets arguing that there is no God, for example—find none. Instead, they find a widespread habit of scoffing. Wits and satirists flourished in the Restoration, making fun of all beliefs, of all claims to authority. Delivering a sermon before the king, Edward Stillingfleet thundered that "no fools are so great as those who laugh themselves into misery, and none so certainly do so, as those who make a mock at sin."

Unfortunately for Stillingfleet, the fountainhead of England's laughter was sitting in front of him, wearing the royal crown. If anyone preferred a joke to a sermon, a mistress to a minister, it was the head of the Church of England. Charles's taste for pleasures was no secret. He had cultivated it in exile in Paris, where he had enjoyed the splendors of the French court, developed a taste for grand wigs, and taken on a string of lovers. When Charles returned to England, he made little effort to hide his nature from the bishops who had consecrated his divine right to rule.

Gilbert Burnet, the bishop of Salisbury, made a careful, unhappy study of his new king. He found that Charles had little real love for people. "He had a very ill opinion both of men and women," Burnet later wrote, "and did not think that there was either sincerity or chastity in the world out of principle, but that some had either the one or the other out of humour or vanity." Despite his distrust, Charles managed to be supremely charming. "He had a softness of temper, that charmed all who came near him," Burnet wrote, "till they found how little they could depend on good looks, kind words, and fair promises; in which he was liberal to excess, because he intended nothing by them, but to rid of importunities and to silence all farther pressing on him."

Charles brought to England a court full of French customs—perfume, face paint, and syphilis. "A strange effeminate age," the diarist Anthony Wood declared, "when men strive to imitate women in their apparel, viz. long periwigs, patches in their faces, painting, short wide breeches like petticoats, muffs, and their clothes highly scented, bedecked with ribbons of all colours." Sexual mixing and matching went on constantly at Charles's court.

Samuel Pepys kept track of the gossip in his journal, describing, for example, a lady-in-waiting who gave birth to a bastard at a royal ball. The king's brother, James, duke of York, was particularly notorious for stealing wives. When a courtier saw him doing nothing more than speaking to his wife, he spirited her away to the country for safety.

Charles, however, was the lecher royal. "His scepter and his Prick are of a length," wrote the earl of Rochester. Soon after Charles's coronation, he snared a new mistress, Barbara Palmer, a twenty-year-old beauty he had first met in exile. When he was married in 1662 to Catherine of Braganza, a Portuguese princess raised in a convent, he made Barbara the queen's lady of the bedchamber. While Catherine proved barren, Barbara soon provided Charles with a second bastard.

As the spiritual leader of the Church of England, Charles left plenty to be desired, going through the motions only to strengthen his political power. He brought back the old ritual of curing the King's Evil, receiving more than twenty-three thousand people at Whitehall in just the first four years of his reign. One by one, they all knelt before him and received his healing touch. Charles was less interested in the holiness of the ritual than in the proof it gave of his divine right to rule. He was indifferent to the bishops' fear of Quakers. As they preached to him, Charles could often be seen dozing in church. William Petty, returned from Ireland, amused the king at court with his satirical impressions of ministers and Jesuits.

Yet for all his sinning and scandal, Charles did not deny God. "He said once to my self, he was no atheist," wrote Bishop Burnet, "but he could not think God would make a man miserable only for taking a little pleasure out of the way."

—

Thomas Willis's secret congregation emerged from Beam Hall at the Restoration and took over Oxford. Ralph Bathurst, his fellow pisse-prophet, became one of the king's chaplains and later was appointed president of Trinity College. Willis's brother-in-law, John Fell, was made bishop of Oxford and dragged the university back to its days under old Archbishop Laud. The sounds of choirs and organs filled the churches once again, while priests revived the sacrament. Fell drummed out Puritans just as harshly as he had been drummed out fourteen years earlier. Some were dragged from their rooms screaming, while those who stayed behind were forced to drop to their knees and recant. No Catholics or dissenters were allowed in the university. John Owen, the Puritan dean of Christ Church, the man in the powdered wig who had looked the other way while Willis hosted his secret congregation, was arrested for holding religious assemblies in his own house. The university took back its control over the city and had Quakers and other dissidents thrown into jail for not swearing allegiance to the king.

At the Restoration, Thomas Willis was thirty-nine years old. For most of his adult life, he had been at war, overtly or covertly, with the rulers of England, and yet he had managed to do well for himself, earning a scientific reputation with his book and Paracelsist drugs. With the return of King Charles, Willis's fortunes climbed sharply, for he was now friends with some of the most powerful men in England, and they were grateful to him. Gilbert Sheldon had endured the siege of Oxford with Willis, serving as the warden of All Souls College until he was ejected by Parliament's Visitors in 1648. Now the new bishop of London rewarded Willis for his loyalty by making him Oxford's professor of natural philosophy. (The Puritan who held the post was fired to make room.)

Fell and the other men who took over Oxford wanted to make the university what it once was, a nursery for the Church of England with the sort of teaching that had gone on when Willis was a boy. Aristotle, one visitor said, "has driven out all other philosophies and rules supreme." But when Willis eagerly climbed to the lectern in the *auctorium* to begin teaching the first generation of Restoration students, he proved a spectacular disappointment to his patrons. Instead of reading from the classics, Willis simply ignored his brother-in-law's rules. Every Wednesday and Saturday morning he described new theories from abroad, the latest work emerging from his circle, and his own starkly anti-Aristotelian ideas on fevers and the body's chemistry. Galen's name was hardly ever heard. If a Puritan had been standing in the hall delivering Willis's lectures, he might have been dragged out, but Willis was a hero of the Restoration, and so he could get away with troubling talk.

Willis believed his new job allowed him to give full scope to his speculations, and he felt obliged, as he later wrote, to "comment on the Offices of the Senses, both external and also internal, and of the Faculties and Affections of the Soul." For years, Willis had pondered the brain, its role in the body, and its function, imagining it as a sort of alembic, into which spirits rose from the blood. Some of his ideas made so much sense to him that he included them in his lectures, but when he heard himself describe them, he realized how flimsy they actually were.

"I seemed to myself like a painter that had delineated the head of a man," he wrote, "not after the form of a master, but at the will of a bold fancy and pencil." It was obvious, he realized, that he "had not followed that which was most true, but what was most convenient, and what was rather desired than what was known. Thinking on these things seriously with myself, I awaked at length sad, as one out of a pleasant dream."[5]

Willis abruptly resolved on a daring project: "to unlock the secret places of man's mind." He would unlock them not by reading Aristotle or Galen but by reading Nature's book. "I determined with myself seriously to enter presently upon a new course, and to rely on this one thing, not to pin my faith on the received opinions of others, nor on the suspicions and guesses of my own mind, but for the future to believe nature and ocular demonstrations."

There was only one course to take. "I addicted myself to the opening of heads."

Willis had seen his share of human brains during his autopsies of dead patients, having made an unprecedented effort to find any damage to the brain that might correspond to the diseases his patients had suffered. But he realized that in order to understand the brain as well as Harvey had understood the heart, he would have to make it his obsession for years. To help him, he enlisted Richard Lower, who supplied his skill with the knife. Willis supplied the brains of his patients. "At length we have had the opportunity of cutting up several," Richard Lower eagerly wrote to Boyle, who was away at the time in London. "And the doctor, finding most parts of the brain imperfectly described, intends to make a whole new draft thereof, with the several uses of the distinct parts, according to his own fancy, seeing few authors speak anything considerable of it."

Beam Hall was transformed from a secret church to a dissecting room. Instead of fugitive royalists, virtuosi now came to watch, listen, and talk. Christopher Wren had been spending time with Willis already, watching him cut open cadavers and drawing what was inside. Now he peered with Willis at brains, speculated on how they worked, and sketched what he saw.

In his fifteen years as a doctor, Willis had performed enough autopsies to know that the classic descriptions of the brain were wrong, as were the usual methods of autopsy, which deformed the brain, slashing vessels and nerves. Willis and Lower dissected the brain in a different way, coming at it from the underside and extracting the brain whole and intact. Willis would then hold it for his audience to see. Looking at the brain this way forced Willis and his friends to think about it in a new light: not as a nondescript mass of flesh glued to the inside of the skull but as an independent organ. With an undistorted view of its intricate structure, they could see for themselves how it was cleanly divided into at least three distinct parts. At the base was the marrow (known today as the medulla oblongata), which joined the brain to the spinal cord. Sitting just above the medulla was the ball-shaped cerebellum, attached to the medulla by a hinge of flesh. Crowning both medulla and cerebellum was the cerebrum, a pair of wrinkled hemispheres. Although earlier anatomists had seen these parts, Willis was the first to demonstrate how distinct they are—distinct enough for Willis to cut each one free and study it on its own.

Earlier anatomists had been far more interested in the ventricles than the brain itself, but when Willis looked at the ventricles, he saw nothing special. They were nothing more than internal gaps formed by the contact between different regions of the brain. The brain itself proved far more impressive to him. By borrowing the tools and tricks of his younger friends, Willis explored it in unprecedented depth. He could immortalize a brain by sinking it in spirit of wine or one of the other preservatives Boyle had discovered, transforming its consistency from butter to boiled egg. Willis was no longer restricted to prodding a brain for a few hours before it rotted. A preserved brain would remain intact day after day as Willis and Lower sliced into it and as Christopher Wren drew its sinuous structures. Willis used the microscopes designed by Wren and Hooke to look at the structure of the nerves, something that no anatomist of the human body had ever done before. He studied how the nerves spread through muscles and organs, forming ever finer branches, which seemed to penetrate every inch of flesh and mingle with equally fine blood vessels.

Willis also borrowed Wren's injections to find the hidden connections and boundaries of the brain. He and Lower opened a dog's skull and injected ink and saffron into one of the two carotid arteries that supplied the dog's brain with blood. They then watched the blood carry the dark stain into its brain. A vast network of fine vessels appeared before their eyes, covering the entire brain. To Willis, they made it "appear like a curious quilted ball."

The closer Willis and his friends looked at the brain's blood supply, the more marvelous it became. The four arteries that carried blood from the neck sent out branches that joined together in a peculiar circle at the base of the brain before entering the brain itself. Willis was among the first anatomists ever to notice the circle; the brain's confusing thicket of fine veins and arteries had hidden it until then. Willis was the only one who realized what it was for.

Staring at the circle, Willis remembered a man he had opened up years earlier. The man had suffered from headaches at one point in his life, but they had gone away, and he had lived many more years. When Willis finally autopsied the man, he found one carotid artery clogged with a yellow, rope-shaped mass while the artery on the other side of his neck was noticeably enlarged. Willis speculated that the man's headaches occurred when the carotid artery became blocked, but the man had not died because the other artery had compensated. Its blood traveled around the circle and flowed into all of its branches, keeping both sides of the brain alive.

To test this idea, Willis and Lower injected a dye into one of the carotid arteries. As Willis had predicted, it stained not one section of the brain but the entire organ. The two physicians tested the idea again with an operation on a spaniel. They opened the dog's neck and tied off all but one of the arteries supplying its brain "very fast and close with silk," as Lower wrote in a letter to Boyle. Then they closed the dog's head again and let it recover from the operation. "The dog was not at all altered by it, but continued very lively and brisk, and was so far from taking unkindly what was done to him that within a quarter of an hour after, he got loose and followed the doctor [Willis] into the town as he visited his patients."

Watching the spaniel trot happily by his side, Willis was ecstatic. He had discovered a wonderfully designed structure in the brain that ensured that the entire brain was supplied by blood, even when part of the structure was damaged. It bears his name today: the Circle of Willis. "Certainly there can be nothing more artificial thought upon, and that can better argue the Providence of the great Creator," Willis declared, "than this fit or convenient disposition of the blood in the brain." As arteries branched off from this network and spread across the brain, they split into twisting microscopic branches—"the serpentine channels of an alembic," as Willis called them. Only after they had entered the brain in this way were the animal spirits—a "chymical elixir," Willis called them—extracted from blood within the brain. Descartes had imagined that animal spirits were particles already present in the blood, but Willis had a different idea: a ferment distilled the spirits within the brain in a chemical transformation.

Just as striking to Willis were the places where the blood did not go. When he injected dye into the arteries, none of it ended up in the ventricles, the cherished chambers where generations of physicians and philosophers had believed the spirits were corralled. Willis could not find a connection by which the spirits could get there from the blood vessels. The ventricles clearly had nothing to do with the business of the mind. They were, Willis wrote, just "a complication of the brain infoldings." The implication was inescapable: Galen and the generations who followed him had mistaken foreground for background.

Willis also found no evidence for Descartes's account of the nervous system, with its spirits pumped through the pineal gland and the ventricles, guided by the wiggling rational soul. The ventricles couldn't hold the spirits as Descartes claimed, nor was the pineal gland special in any way to humans, since Willis could find it in the brains of birds, even of fish. If the pineal gland had a function in these animals, "we can scarce believe this to be the seat of the Soul," Willis wrote. The ventricles are a "mere vacuity," about which there was "no reason we have to discourse much."

Abandoning the ventricles, Willis followed the blood instead, deep into the flesh of the brain. Seeing that the arteries split off into finer and finer vessels that spread over the brain's entire surface, he found that each major part of the brain was supplied by its own set of vessels. Here was more evidence that the brain was made of distinct parts that carried out different jobs, like the parts in a clock. As Willis and Lower sliced into the brain, unfolding pieces so that they lay flat on a dissecting table, they discovered it had far more parts than had ever been imagined. Earlier anatomists had divided the cerebrum into two regions—an outer bark, or cortex, made of gray matter, and an inner core of white matter. Willis discovered gray structures characterized by a pattern of stripes buried deep in the cerebrum that traveled up from the brain stem and into the higher reaches of the brain. He called this region the corpus striatum (the striped body).

Willis and his team also followed the nervous system down from the brain and through the entire body. They traced the cranial nerves that sneaked through holes in the sides of the skull and wandered their way to the tongue, the voice box, the lips, the teeth, and the cheeks. They followed the vagus nerve, traveling along its fan of fine branches that entwined the heart and diaphragm. The virtuosi gently teased apart bundles of fibers into their individual branches and marked where each one ended. With unmatched precision, they drew giant diagrams that looked like topographical maps of streams and rivers and watersheds.

For two years, from 1661 to 1663, Willis and his company created a gory mess. "No day almost passed over without some anatomical administration," Willis wrote, "so that in a short space there was nothing of the brain and its appendix within the skull, that seemed not plainly detected and intimately beheld." Although human brains were most precious to Willis's team, they also learned much from the brains of animals. Willis would later recall how they had "slain so many Victims, whole Hecatombs almost of all animals, in the Anatomical Court."

Willis carried Harvey's methods from blood to brain. By comparing the brains of humans and animals, he found that the brain of a fish or a cow had the same basic architecture as a human's—a medulla, cerebellum, and cerebrum. Compared to animals, however, humans had a gigantic cerebrum, gnarled with convolutions. Willis believed that the difference in shape must mean a difference in how the brains functioned. Humans and animals had practically identical cerebellums covered by a regular pattern of ridges and streaked inside by a starburst pattern. The cerebellum's simple texture suggested that it worked like a simple machine. It created spirits that traveled down to the heart and other organs and kept them in clocklike motion without any supervision of the higher faculties. "The Spirits inhabiting the Cerebel perform unperceivedly and silently their works of Nature," Willis later wrote, "without our knowledge or care."

As Willis studied the cerebellum, he thought back to autopsies he had made of his patients, recalling that people with pains in the back of their head, where the cerebellum was located, often suffered "cruel and horrid symptoms" in their lungs and heart. To prove that the cerebellum controlled those organs, he and Lower opened a living dog's chest and tied the nerves that ran from it to the heart. The chambers of the dog's heart became engorged with blood and the animal quickly died.

This was a revolutionary idea in two ways. Aristotle had divided movements into involuntary and voluntary ones, but Willis now assigned them to different parts of the nervous system for the first time and even carried out experiments to back up his claim. It was also revolutionary in the way it drained the soul out of the heart. Willis argued that the heart was tethered to the brain, following the commands of the spirits sent down to it from above. Now a mere muscle, robbed of Galen's vital soul and natural intelligence, the heart was no longer the moral center of Christianity, no longer king of the body. Willis handed that title to the brain.

Willis still had to account for how the king ruled his kingdom. "Some think it is enough to say, that the Soul itself, by its presence, doth actuate the Muscle," Willis wrote, but as a mechanical philosopher, he found this kind of reasoning both supernatural and useless. Why then couldn't the soul "bend and force heavy and very great bodies whither it pleases?" The nervous system and the body that contained it were both machines, Willis believed, and so it should be possible to "explicate them according to the Rules, Canons, and Laws of a Mechanick."

Descartes had imagined that muscles moved hydraulically. Spirits were pumped by the heart into the ventricles of the head and from there through hollow nerves. When they reached the muscles, they inflated them by sheer force. But Descartes had never examined nerves closely. When Willis and his friends looked at them through a microscope they saw solid cords with small pores, like sugar cane. Reaching back to his earliest days of alchemy, Willis found a new way to account for how this sort of nerve could make a body move. He envisioned a nervous juice flowing through the nerves and animal spirits riding it like ripples of light. The spirits did not move muscles by brute force but rather carried commands from the brain to the muscles, which responded with a minuscule explosion. Each explosion, Willis imagined, made a muscle inflate.

Descartes had envisioned the spirits of the body driven ultimately by the beating heart, which pushed them throughout the brain and into the muscles. Willis reorganized the body, making the brain the origin and the nervous system an explosive fountain. Just as the brain sent out spirits to the heart and the other organs, the nerves picked up signals from the outside world and sent animal spirits flashing back into the brain. The spirits carrying these impressions raced along pathways that led deep into the flesh of the brain to the striped body (the corpus striatum). Willis imagined the corpus striatum acting like a lens, focusing the spirits and projecting them onto a white expanse of flesh that joined the brain's two hemispheres, known as the corpus callosum. It served as a central meeting place for the spirits of the brain, a mental market of sorts. Some spirits were reflected from the corpus callosum through the cerebellum back into the body, triggering an involuntary reflex. (Willis was the first writer to use the word "reflex" to describe this kind of automatic reaction.) This quick bounce of the spirits was responsible for making a soldier jump at the sound of a cannon.

In other cases spirits ascended beyond the corpus callosum and struck the cortex, where Willis believed the higher faculties resided. With enough force, the spirits could even make a permanent impression on the cortex, creating a memory. Here in the cortex the circulating spirits gave rise to imagination, appetites, and even reasoning.

In order to produce these complex thoughts, the spirits of the brain had to take complex paths. Willis claimed that they followed the winding furrows of the cortex. Was it any surprise, then, that a bird had more of these furrows than a fish and a cat more than a bird—or that humans had a maze far beyond anything found in any animal's brain? "These folds or rollings about are far more and greater in a man than in any other living Creature," he wrote, "to wit, for the various and manifold actings of the superior Faculties."

Descartes was the first philosopher to imagine the human body as nothing more than the union of a rational soul and an earthen machine, but Willis was now backing up a mechanical account of thought with an accurate anatomy of the brain. In man and fish alike, he argued, spirits travel through the flesh of the brain and out into the body to produce involuntary reflexes. Willis also found a way to account for the mysterious sympathy between the organs of the body.

He rejected van Helmont's mystical influences, arguing instead that they were joined by interconnected branches of the nerves that emerged from the cerebellum. One web of nerves spanned the intestines, creating orderly waves of peristalsis to carry food through them. Branches of the vagus nerve joined the intestines and liver, so that the liver could inject the right amount of bile to help the bowels digest food. Laughter, Willis reasoned, made us smile, wrinkle our eyes, and take short breaths because the same nerve sent branches to the mouth, eyes, and lungs. Nerves running through the entire face joined together in a single bundle of nerves near the base of the brain, bringing a consensus to the chin, cheek, eyes, and mouth. The kiss of lovers immediately fired up their loins because of another branch shared between mouth and genitals. Willis replaced the occult mystery of sympathy with a harmonious network of nerves and showed how they worked as a system—something no one had done before.

By the spring of 1663, Willis's team had finished their explorations. Lower drew up diagrams of the nervous system, while Christopher Wren drafted illustrations of the brain itself. Wren's beautiful drawings revealed the brain as a delicate, complex organ with the beauty of an orchid. He managed to produce these masterpieces despite a flood of work that was beginning to submerge him. The previous year, Wren had designed a small chapel for his uncle Matthew, freshly sprung from the Tower of London and reinstated as bishop. His skills in geometry, physics, and drawing served him well as an architect. Gilbert Sheldon, who had just been appointed archbishop of Canterbury, heard about Wren's work and asked him to design a theater. It would be Sheldon's gift to Oxford, a majestic building where degrees would be conferred. In the spring of 1663, Wren was designing his first major building, the magnificent Sheldonian Theatre. His pen hopped from building plans to the brain and back.

Willis synthesized everything he and his friends had found, salting their raw observations with his own speculations about the movement of spirits through the brain and nerves. He created what he called a "neurologie," a "doctrine of the nerves." His old friend Ralph Bathurst polished Willis's Latin, and in 1664 he published a book entitled *Cerebri anatome,* or *The Anatomy of the Brain and Nerves.*

Only five years had passed since Willis published his book on fevers and ferment, but the business of spreading scientific knowledge had changed dramatically. The intellectual free-for-all under Cromwell was over. Willis now had to get a license from the archbishop's censors for each book he hoped to publish. And in place of the invisible social networks through which the new science traveled in the 1650s, an institution had sprung up to formalize the spread of the work of men like Willis: the Royal Society for Promoting Natural Knowledge.

The Royal Society was born in London on a November afternoon in 1660. Christopher Wren had just given a lecture at Gresham College and retired with some other virtuosi to a scholar's rooms at the college for a drink. John Wilkins was there; he had left Oxford for Cambridge, but the fall of the Puritans had forced him from his post. William Petty was there as well and likewise in limbo, having fallen out of the graces of the men put in charge of Ireland after Oliver Cromwell's death. As unsettled as their own lives might be, the virtuosi all believed that the time might be right to create an official organization for their experimental philosophy. Wilkins, Petty, and others had been talking for years about re-creating the Oxford circle on a grander scale. On that November day, the dozen virtuosi all agreed that such a thing should exist. Wren scribbled down a list of forty names of potential members to ask, including Willis, Hooke, and Boyle.

The Royal Society began to meet every Wednesday at Gresham College to watch experiments, discuss them, and hear reports of wonders from abroad. Royalists, Puritans, and even Catholics were all welcomed as fellows of the society. In place of pamphlets and haphazard letters, the society's secretary, Henry Oldenburg, published a regular journal and distributed it throughout England and much of Europe. The founders also approached Charles II, promising him that the virtuosi worked "for the riches and ornament of our kingdom."

Whereas Francis Bacon had failed to entice Elizabeth I and James I with his proposals, a few decades later, the virtuosi discovered that Charles was fond of the new science. On his return from exile the king had brought a personal alchemist and set up a private laboratory; he also enjoyed watching anatomists dissect human cadavers and even dissected a few animals himself.

He gave the virtuosi a royal charter, but they had to pay for it in a currency of amusement and hard labor. Christopher Wren presented the king with a three-dimensional map of the moon, a painted, carved pasteboard globe complete with mountains and valleys. The king would come to meetings of the society to be entertained by Boyle with his remarkable air pump. He demanded an artificial eye be made but never bothered to come see it.

The Royal Society claimed to be an ally of both the king and the Church of England, although it knew that not every priest would believe it. Suspicions circulated through the church that the virtuosi did not have the proper Protestant outlook. Bishop Thomas Barlow declared that "it is certain this New-Philosophy (as they call it) was set on foot, and has been carried on by the ants of Rome, and those whose oath and interest is to maintain all her superstition."

The Royal Society produced a steady stream of propaganda to calm conservative nerves, smoothing over the embarrassing details. The official history of the Royal Society described how it grew out of the Oxford circle but skipped over the fact that the circle had taken shape thanks to the purges and appointments of king-killing Cromwell. Instead, it described how a band of gentlemen was drawn to Oxford by the tranquillity of the place. "Their first purpose was no more than only the satisfaction of breathing a freer air, and of conversing in quiet with one another; without being engaged in the passions and madness of that dismal age."

The Royal Society also distanced itself officially from alchemy, which it claimed to be little different from the delusions of religious fanatics. Yet a number of their members were alchemists who hoped to find the philosopher's stone, including Robert Boyle, now England's most famous natural philosopher. Boyle himself helped to calm the church's fears about his work, preferring the harmless-sounding word "corpuscle" to "atom," which might have linked him with Epicurus. "When I speak of the corpuscular or mechanical philosophy," Boyle wrote, "I am far from meaning with the Epicureans that atoms, meeting together by chance in an infinitude vacuum, are able of themselves to produce the world."

The virtuosi claimed that they investigated nature by the light of reason. By showing the nation what they discovered, they promised to make God's work clear and provide an antidote to fanaticism. The official history of the Royal Society claimed that natural philosophers are "truly acquainted with the tempers of men's bodies, the composition of their blood, and the power of fancy," which allows them to tell "the differences between diseases and inspiration." Boyle and other members of the Royal Society spoke and wrote endlessly about how their natural philosophy was the best weapon that the church could hope for, even encouraging members to gather evidence of witchcraft to destroy the arguments of atheists.

The Royal Society searched for ways to show off the value of its work. Robert Hooke was appointed curator of experiments and presented the society with a new experiment every week. When the virtuosi saw things that no one else could see, they illustrated them to make the world their witness. Christopher Wren presented Charles with some of the marvelous drawings of insects that he had made with the help of a microscope in Oxford in the late 1650s. The Royal Society and the king agreed that Wren should put together an entire book of pictures, but he wriggled gracefully out of the colossal task, and his old partner in microscopy, Robert Hooke, agreed to carry on the project.

Micrographia, the book that Hooke produced over the next three years, was a jolting revelation. Until then the microscopic universe had been a secret game park, with admission limited to only a few natural philosophers. Now Hooke offered a tour, one that he crafted as pictorial propaganda for the new science. The first image his readers encountered was of the simplest thing imaginable, the point of a needle. Under Hooke's microscope, it was transformed into a craggy mountain. A body louse sprawled over one of Hooke's foldouts, its anatomy as intricate as any to be found in a bird or a flower. A fly's eyes were resolved into thousands of carefully laid tiles. A smooth stem of a plant was magnified into millions of juice-filled cells. God's handiwork extended down to the microscopic scale—clear evidence, Hooke wrote, that "Nature does not only work mechanically, but by such excellent and most compendious as well as stupendous contrivances, that it were impossible for all the reason in the world to find out any contrivance to the same that should have more convenient properties." He promised even greater things from microscopes in the future. "We may perhaps be enabled to discover all the secret workings of Nature, almost in the same manner as we do those that are the productions of Art, and are manag'd by Wheels, and Engines and Springs."

The Royal Society demonstrated that God's work could be witnessed in the leg of a flea. It could also be seen in the folds of the brain. In 1664, the same year that Hooke completed *Micrographia,* Ralph Bathurst came to the Royal Society bearing a copy of Willis's *The Anatomy of the Brain and Nerves*. (Willis, it seems, was either too shy or too busy with patients to come himself.) Bathurst presented the society with the book, in which the members could see the divine workmanship hidden in their skulls.

Unlike Harvey, Willis did not have to wait until he was old and weak to see his fame spread. Pamphleteers saluted him as "the ornament of our nation, next to immortal Harvey." His fellow anatomists referred to *The Anatomy of the Brain,* the first work ever dedicated completely to the nervous system, as Willis's "immortal book on the brain." Four editions were printed in a single year, and anatomists across Europe were soon carrying a pocket-sized version with them. Willis had mapped the brain just as Hooke had mapped the microscopic world and revealed it to be an equally marvelous realm. The ingenious loop of arteries that supplied the brain became known as the Circle of Willis, while Wren's drawings proved so accurate that they were still being reproduced in textbooks in the twentieth century. All told, *The Anatomy of the Brain* would go through twenty-three editions, and well into the nineteenth century it would be required reading for anyone who would call himself an expert on the brain.

Willis's team had produced more than a map. They had, for the first time, created a unified treatment of the brain and the nerves. An erratic, error-ridden study of the brain became a rigorous, experimental science, to which Willis gave the name neurologie. For his efforts, one twentieth-century neuroscientist declared Willis "the Harvey of the nervous system."

The Anatomy of the Brain was the first of a trilogy of books by Willis, each of which would help clear away old notions of the brain's workings and establish many new ones that still dominate our thinking today. Only someone as full of contradictions as Willis could have embarked on such an enterprise—a physician skilled in both anatomy and alchemy, who had embraced the new mechanical philosophy as well as van Helmont's mystical ferments; a respectable figure nestled at the very heart of the Restoration establishment, a soldier for the king's murdered father, and an Oxford professor appointed by the archbishop of Canterbury.

For all his respectability, though, Willis had to make it clear that *The Anatomy of the Brain* wasn't the work of an atheist. Under the protection of none other than Archbishop Sheldon, to whom he dedicated his book, he wrote, "Once more your Sidley professor and your servant (the more happy title), flings himself at Your feet, with this only ambition, that he might render something of thanks for Your kindness and benefit." Willis wrote that it only made sense that an investigation such as his, which was designed to "look into the living and breathing chapel of the Deity," should be dedicated to Sheldon, "who most happily presides (both by merit and authority) over all our temple and sacred things." The animals slain in Willis's anatomical court must rightly be brought "to the most holy altar of Your Grace." Even when Willis filled Beam Hall with corpses, it remained his personal church.

Chapter Nine

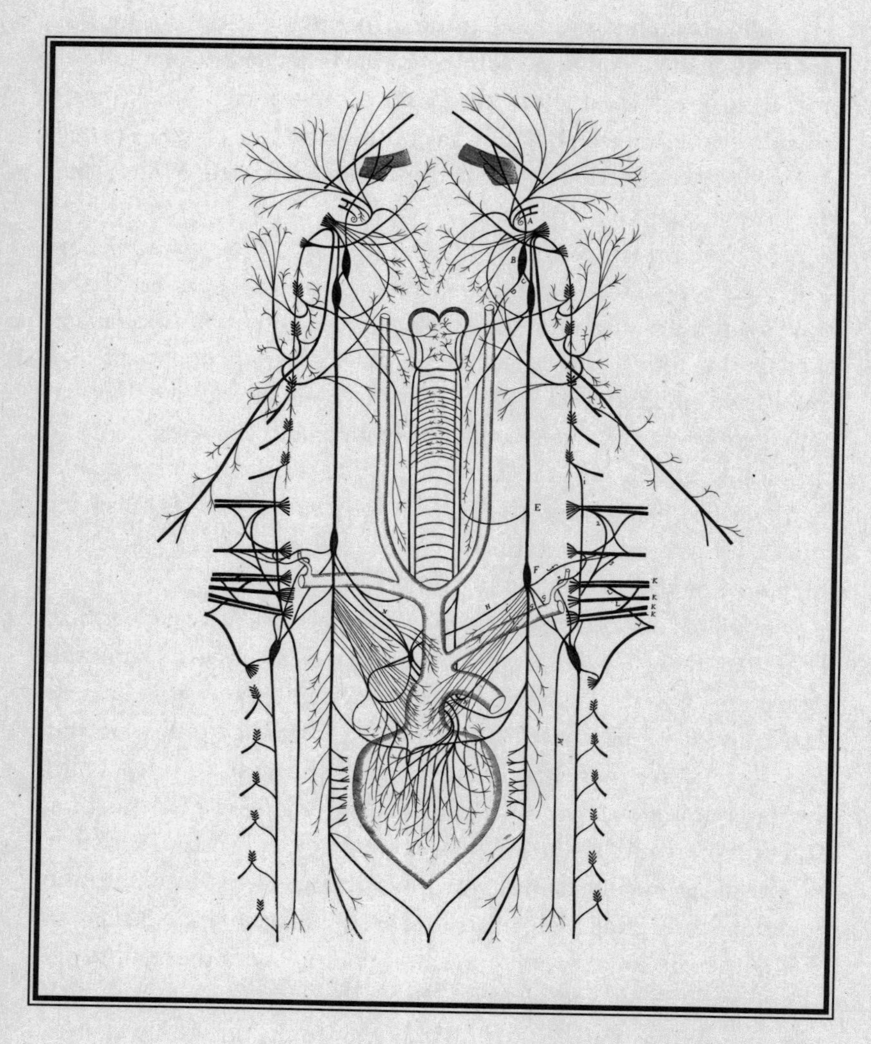

A DIAGRAM OF THE SYMPATHETIC NERVES, FROM *THE ANATOMY OF THE BRAIN AND NERVES*. IN PLACE OF MYSTICAL UNIONS BETWEEN THE HEART AND MIND, WILLIS TRACED A NETWORK OF NERVES.

Convulsions

Like his book on fevers five years earlier, *The Anatomy of the Brain and Nerves* brought Willis more business. In 1665, a nobleman wrote to a friend that Willis was "one of the learnedest and most famed physicians in the world." Along with another Oxford physician and a surgeon, Willis bought the Angel Inn on High Street. They turned the tavern into what amounted to a hospital in all but name, complete with a "fluxing chair" for sweating out syphilis. Wealthy invalids riding from London to Bath stopped off at the inn for treatment. Now in his mid-forties, Willis had become the richest man in the county of Oxfordshire. He went on treating poor patients, but out of charity rather than necessity. He no longer traveled to the markets in the surrounding towns as a pisse-prophet, nor did he ride for hours on a shared horse to visit a sick farmer who might or might not be able to pay him. When Willis left Oxford, it was to tend to nobility.

Not long after he published *The Anatomy of the Brain,* Willis set out on a journey north from Oxford to Warwickshire to see one of the noblest women in all England. He traveled down remote roads that led to a vast estate owned by Lord Conway, secretary of state to Charles II. His carriage pulled up before Ragley Hall, Conway's majestic mansion.

Willis was guided down its long corridors to a darkened room. Lying in a bed before him was Lord Conway's wife, Anne. She was a beautiful, dark-eyed woman in her early thirties, who lived in almost perpetual pain.

Willis had been summoned to cure Lady Conway of headaches so powerful that even a glimmer of light sent her into agony. He examined her and asked her about her life, listening carefully and compassionately. Out of these house calls came the first clinical descriptions of migraine in the history of medicine, but these meetings were also important for another reason: they were a historic encounter between two of the seventeenth century's most radically different visions of the soul. Thomas Willis was in the process of installing the soul in the brain, which he saw as part machine, part alembic. He was about to take his neurology another great step forward, using it to illuminate what happened when the brain and nerves became diseased, as in the case of Lady Conway. Illnesses that had been put down to imbalances of the body's humors or even demonic possession Willis now conceived as the wayward actions of mechanical particles. Lady Conway, meanwhile, was a brilliant philosopher in her own right who spent years probing the nature of the soul. For her, the headaches that plagued her were not just a matter of errant particles, but were a sign of her union with an entire universe of spirit.

Anne Conway probably had the best education of any Englishwoman in her generation. "She was of a most beautiful form, and a great wit," Willis later wrote, "so that she was skilled in the Liberal Arts, and in all sorts of Literature, beyond the condition of her sex." Unable to go to a university, she educated herself, learning Latin and French and reading philosophy before the age of twenty. She traded a steady stream of letters with her brother John while he was at Cambridge, keeping up to date with the newest philosophies circulating there. Before long she wound up in a correspondence course of sorts with her brother's mentor, the philosopher Henry More.

It was More who called the brain a bowl of curds incapable of carrying out the workings of the spirit. He did not have a low opinion of the brain so much as a high opinion of the soul, which could not possibly be accommodated in such a mushy organ. More had built his life's work around defending the soul—and the rest of the spirit world—from the evil twins of materialism and atheism. Influenced more by Plato than by Aristotle, he believed the immaterial soul occupied the body as a temporary vessel. The soul had existed before the body and would go on existing afterward as part of a busy community of spirits.

More first read Descartes in 1645 and was instrumental in bringing his ideas into England in the 1640s. At the time, More thought Descartes had found a way of understanding the world that had been prophesied in the Bible. "I should look upon Des-Cartes as a man more truly inspired in the knowledge of Nature than any that have professed themselves lo these sixteen hundred years," he wrote. Whereas Petty complained that Descartes had no experiments to back up his claims, More saw experiments merely as demonstrations of what reason discovered. More believed that when the philosophies of Descartes and Plato were finally fused together, all religious wars would stop, the barrier between the physical and spiritual worlds would dissolve, and a millennium of peace would arrive.

After a few years, however, More's enchantment began to fade. Descartes held that the soul did not take up space, that is, it was not "extended." More could not see how such a soul could exert any physical force on its material body. At the same time, dead matter seemed too feeble to move other matter at a distance. How could a magnet pull iron to itself, he asked, or a plucked string make a neighboring one quiver in sympathy? More came to believe that only spirit could make the universe move, that an object was roused to motion like a mind to a thought. "The phenomena of the world cannot be solved merely mechanically," he declared.

In order to move matter, More argued, spirit must be extended like matter. Souls occupied space. Unlike matter, though, they could change their shape and penetrate material objects. Individual human souls and other residents of the spirit world—ghosts, demons, angels, and even God—were extended. God was infinitely huge and encompassed everything in the universe. Space was the mind of God, More claimed, which contained His thoughts. A far cry from the business-minded experiments at the Royal Society, More's philosophy was mystic poetry. "I brush the stars and make them shine more bright," he declared.

Anne Conway had her brother pass a letter to More asking for help in understanding Descartes. She wanted to know, for example, how there could be a fully perfect being—God—without a fully imperfect being existing as well. When she was not satisfied with his explanations (how for example, the green in a green carpet did not exist in the carpet but in the soul of the person looking at the carpet), she demanded a better explanation. Soon Conway and More were in deep discussions of philosophy, including More's own work. In an age when men generally considered women too weak-minded for such stuff, More wrote to Conway without a drop of condescension. In 1652 he dedicated his book, *An Antidote Against Atheism,* to her, writing that "you have not only outgone all of your own sex, even of that other also, whose ages have not given them over-much the start of you."

Anne achieved all this despite years of pain. At age twelve, she suffered a fever, which was followed by a bout of crippling headaches that returned every few months. They grew worse in her twenties, making her vomit and forcing her into darkened rooms for days on end. Her family did not know quite what to make of her. Despite her spells of pain, she was charming, fearless, and obsessed with science. She taught herself Greek and read

Euclid. When she tried to convince her father-in-law that Copernicus's solar system was not just a fancy, he told her, "You write like a man." She welcomed Henry More to her house for months at a stretch, along with other philosophers and theologians, to debate the nature of matter, souls, and God. Her family assumed that her headaches were caused by all her unnatural thinking.

Over the next decade a long column of physicians marched its way through her family mansion. Her first doctor was none other than William Harvey, whose niece had married into Anne's family. Now in his seventies, aching from gout, bitter at the collapse of his career, Harvey was not much help to Anne. While her family knew of Harvey's unusual ideas about the body, they weren't impressed with him as a doctor. Her father-in-law warned Anne that "to have a Physician abound in phantasie is a very perilous thing." Anne found Harvey more preoccupied with his own illnesses than hers. ("Yet he pretends very much to study and lay my case to heart," she wrote.) Harvey acted as Conway's doctor for at least two years, but the only recorded suggestion that he made to his young patient was that she should have a hole drilled into her skull.

Anne found new doctors. Some offered her opium, some applied plasters of lead and soap to the back of her head. Theodore de Mayerne, the champion of Paracelsus at the royal court, gave her drops of mercury that nearly killed her. Others poured drops of water on her head, prescribed new drugs such as tobacco and coffee. She even traveled to France to have her head drilled but at the last minute changed her mind. The French doctors bled her jugular vein instead, bringing her no relief. Anne returned to England still submerged in pain. When she became pregnant, she hoped that the birth of her child would heal her, but her son contracted smallpox before his second birthday. Anne refused to be parted from the boy until he died, and she ended up catching smallpox from him. It took months for her to recover from the infection, unable to tell where her pain stopped and her grief began. "It hath pleased God to exercise me by divers afflictions and by one so sensible in the death of my child," she wrote.

Her husband took Anne to Ireland in order that she might escape her suffering. Anne loved the Conway estate, with its two-thousand-acre park and the dozen islands peering out of the loughs nearby, but by 1664 her headaches had blurred together into an unbroken stretch of pain. Edward Conway brought her back to England. When they stopped at Ragley Hall on their way to London, Anne declared she was too exhausted to move farther. "I cannot dissemble as much as not to profess myself very weary of this condition," she wrote. She would not leave Ragley Hall for the rest of her life.

She continued to seek help. Robert Boyle recommended that she take essence of copper, the compound he and Starkey had created, a cure-all that he claimed he used to heal hundreds of children of rickets. She drank it in Canary wine and in sack, but it helped her neither time. An Irish healer named Valentine Greatrakes came to Ragley Hall promising to cure her with the stroke of his hand. He held her head for days, his skin smelling of flowers, and yet her pain did not fade.

Thomas Willis had encountered Anne's kind of pain before. It was "constant to no temperament, constitution, or manner of living, nor to no kind of evident or adjoyning causes," he wrote. It "ordinarily falls upon cold and hot, sober and intemperate, the empty and the full-bellied, the fat and the lean, the young and the old; yea upon Men and Women of every age, state, or condition."

Willis visited Anne regularly, but as the months passed, she began once again to lose hope. In a letter to her husband, she wrote, "I shall not send to Dr. Willis this week, because I have but newly begun the trial of his last prescription, and therefore cannot give him any account, but have little hopes it should succeed better than the rest have done."

In the end, Willis failed to cure Anne, just as all the others had failed, but he distinguished himself in his keen observation and gentle care. He was appalled, for example, that she had been given mercury. "Having tried that kind of remedy in headaches arising from other causes," he wrote, "I found not the harvest worth the pains, and I confess some examples of those kind of cases, have terrified me from that method."

Anne Conway stopped sending for Willis around 1666. She had lost all hope in medicine, in the bleedings and the strokings, the plasters and the tobacco, in the authority of Galen and the new claims of the virtuosi. For the next four years she simply held her own against the pain. Her hope rose again only in 1670, when she received a letter from Henry More with astonishing news. Franciscus Mercurius van Helmont, the son of Joan Baptista van Helmont, had arrived in London. After his father's death, van Helmont had wandered the Continent and had gained a reputation as a mystical healer. For years Anne Conway had wondered if he might succeed where others had failed. More persuaded van Helmont to visit Ragley Hall. Once the doctor met Lady Conway, he lost his wanderlust and settled in at the mansion. At first Anne Conway felt better under van Helmont's care, but before long her headaches returned. He stayed on, though, and gave her a different sort of comfort: rather than eliminating her pain, he helped her accommodate herself to it.

Pain is the one unbroken thread throughout van Helmont's own mysterious life. Historians have few details that they can use to reconstruct it beyond his own mythology. In his books, he boasts that he battled angry mobs single-handed, fought off bandits, rescued forty horses from a burning barn, and survived shipwrecks. Rumors gave him all sorts of magical powers, even suggesting that he owned the philosopher's stone. What is clear is that after van Helmont published his father's book, he began to wander Europe. He worked as a painter, a weaver, a chemist, and a physician. He lived with gypsies and had princes for patients. Throughout all that time he suffered pain. It began when a tree van Helmont ordered be brought down crashed on top of him. He became convinced that the spirit of the tree was responsible for his wound, and he had a salve made from its sawdust, which he applied to his broken shoulder. It did no good, and he was left in agony. He had no alternative but to explore the pain.

"Having made a thorough anatomy of pain in all its parts," he wrote, "I found that pain was nothing but my own life, excited or inflamed for my own good. I began to love the pain." Within a few days he was cured of the injury. The mind was so intimately linked with the body, he claimed, that it could heal any disorder by embracing its pain. In 1661 van Helmont was jailed by the Inquisition for disparaging Catholicism. He looked forward to the rack, saying that "whatsoever punishment happens to a man that is innocent, it doth him much good." Through pain the soul was regenerated and purified and thus better able to join with God.

Anne Conway had never met a doctor quite like van Helmont. "I have some relief (God be thanked) from his medicines, I am sure more than I ever had from ye endeavours of any person whatsoever else," she wrote her brother-in-law, "but yet I have had much more satisfaction in his company." Van Helmont showed her how to find a meaning to her pain beyond an imbalance of humors or a disturbance of particles. It became an organizing principle that let her make sense of the philosophical questions she had been asking for thirty years. In his company, she felt so inspired that she began to write an essay, laying out a philosophy of her own.

Descartes had drawn a bright line between the rational soul and the rest of the universe, which was dead matter. Anne Conway believed her life of pain had destroyed that dividing line. In her essay, she wrote of the pain a soul can feel—a searing headache, for example—in response to certain experiences of the body. "Why does the spirit or soul suffer so with bodily pain?" she asked. "For if, when united to the body it has no corporeality or bodily nature, why is it wounded or grieved when body is wounded, whose nature is so different?"

Lady Conway explained her pain by interpreting the world in a new way. She eliminated "dull and stupid matter" from the universe and in its place put spirit alone. Everything in the world was united. Not all spirits were alike, of course; they were arrayed along an unbroken continuum from crude to perfect, from boulders to animals to humans to angels to Jesus to God. Our bodies, she said, were a kind of congealed spirit, but no matter how crude something might be, even a chunk of ice, it was not cast out of the spirit world. It still had a living force within it and had a capacity for change, for transformation to a more perfect form, as when ice turns to liquid and finally to gas. Our own bodies went through this perfection as they turned food into animal spirits, which she called "the proper angels of man"—angels, of course, being divine messengers.

The more corporeal something was, the more pain it felt, but suffering also brought about a regeneration that raised a being up the scale of spirit through cycles of life and death. Ultimately, death was an illusion. "For how can a dead thing depend on him who is life and charity?" she asked. Religion—any genuine religion, to Lady Conway's mind—helped human souls rise up the scale of perfection toward God. "If it be granted that the soul is of one nature and substance with the body, although it is many degrees more excellent in regard of life and spirituality," she wrote, "then all the aforesaid difficulties will vanish."

Anne Conway apparently never told anyone about her essay. She simply put down her black pencil and tucked her notebook away. Now in her forties, she had only a few years left to her life—her "exclusion from the world," as she called it. She spent most of that time in the company of Quakers, whose quiet patience in the face of their suffering impressed her. On one of his trips back to Ragley Hall, Lord Conway was appalled to see Anne surrounded by "an unpleasing sort of people," as he wrote to a friend, "silent, sullen, and of a removed conversation." Henry More discovered that his pupil no longer needed him. He stayed at Ragley Hall for the entire summer of 1677, but saw Anne only once or twice.

Anne did not care what anyone thought. Within her dark room, she was queen. "The weight of my affliction lies so very heavy upon me, that it is incredible how very seldom I can endure anyone in my chamber," she wrote to More, "but I find them [the

Quakers] so still and very serious that the company of such of them as I have hitherto seen, will be acceptable to me, as long as I am capable." Their suffering and their faith brought her closer to God than "the most learned and Rhetorical discourses of resignation can do," she wrote. In a sense, the Quakers had returned to the sort of psychological medicine physicians had practiced a century earlier in England and much of Europe, a mixture of spiritual counsel and prayer. Anne Conway converted to Quakerism, and when death came for her soon afterward, she faced it peacefully.

Before van Helmont left Ragley Hall to take up his wandering again, he performed a final service for her. He built a wooden coffin lined with pitch and fitted with a glass lid, and he filled it with spirit of wine. Anne's body floated in it while her husband made his way back to Ragley Hall, until he could look upon her face one last time, sealed in a preserved case for observation like one of Willis's brains. Afterward her coffin was placed in a lead casket and interred, as she had wished, without ceremony or ornament in the church in the village of Arrow.

Van Helmont discovered her essay after her death and arranged for it to be published as a book entitled *The Principles of the Most Ancient and Modern Philosophy.* It influenced thinkers as important as Gottfried Leibniz in the years after her death and lived on in the counterphilosophies of the soul that sprang up in opposition to the mechanical, chemical tradition Willis helped establish. But for many years the world forgot about Lady Conway. Her name, like those of most women writers of the seventeenth century, did not appear on the title page of her book, and in time people came to assume that van Helmont had written it. All that remained to mark her life was a scrawl of graffiti someone left on her coffin: "Quaker Lady." She would have asked for nothing more.

—

Lady Conway was a puzzle to Willis, one that he continued to mull after his service to her. Having erected a new anatomy of the brain and nerves, he used it to try to understand disorders such as her headaches. He was not satisfied with the old accounts based on Galen, stories of vapors trapped in the skull and imbalances of black bile. Willis looked for an explanation consistent with his own philosophy, reflecting on the headaches his other patients had suffered over the years and what he had seen when he had opened their skulls after death. His accounts of their headaches were more detailed and accurate than any in the history of medicine—the way a spasm of pain can creep across the head, for example, or the way a wolfish appetite in the evening can foretell a migraine attack the next day. To account for these pains, he envisioned the nerves that lined the skull as being vulnerable to irritating substances. Their nervous liquor might stagnate, making the nerves swell. If a patient's blood was then stirred up, it might rush into the blood vessels in the head and pull apart the swollen nerves, creating pain.

Willis was speculating, of course, but in a new way, looking for explanations based on the movement of particles in the blood and nervous system. His speculations went beyond headaches to the madmen he had treated, to the fools, hysterics, and convulsives, to people who slept for days on end and to people who got no sleep at all. Willis set out to write another book on the diseases of the brain.

One night in December 1664, as Willis was mulling his new project, he happened to look up at the sky. A strange new star hung on Orion's belt, big and streaked. It was, John Wallis informed him later, a comet. A second comet appeared in March. The first was heavy and solemn, the second swift and blazing. Christopher Wren and Robert Hooke trained a telescope on them to calculate their paths and determine whether they were the same brilliant body just before and after it swung behind the sun. Astrologers saw in the comets something else: a warning of plague and fire. Indeed, in April 1665, reports began to spread out of London that pestilence had arrived.

For centuries, the plague had paid terrifying visits to England.

No other disease could kill so many people so quickly, as fleas hopped from infected rats to humans. The bacteria they transmitted in their bites brought raging fevers, hideously swollen lymph nodes, and in many cases a swift death. At the end of 1664, ships from Turkey brought the plague to the Netherlands, and a few months later, Dutch smugglers brought it to London.

At first, the plague affected only a few victims in the slums, but old people remembered how the previous plague, forty years earlier, had gained strength with the spring. As temperatures rose and rats and fleas thrived, the plague spread through the cramped streets. Soon so many people were dying that the city began to shut up the houses of the sick, marking a cross on the door and posting guards to keep them from leaving until they recovered or died. Old women trolled through the street to count the dead, and each week parish clerks tallied the totals. At first hundreds died each week, then thousands. A stream of Londoners escaped into the countryside in wagons, on ferries, or on foot.

"Death, as it were, rode triumphant through every street," wrote one chaplain, "as if it would have given no quarter to any of mankind, and ravaged as if it would have swallowed all mortality." Coffins were the only things for sale in the city, and by the end of summer the dead were being dumped from carts into giant pits. Families trapped in their houses murdered the guards watching over them. Robbers rubbed themselves with spiced vinegar before they broke into houses to pick over corpses. The wealthy physicians fled the city, leaving it to the medical alchemists, who believed that at last they would be able to prove their superiority to the Galenists. But their mercury drinks and toad amulets failed them.

From the safety of Oxford, Willis sent some notes to Archbishop Sheldon in London about how to defend against the plague. Willis recommended smoking tobacco, noting that in the previous plague no tobacco shop was infected. To "ripen" the sores of plague victims, he recommended poultices made of onion and white lily roots. But he was open to other possibilities, perhaps because they were all so equally useless. "Some recommend live frogs to be apply'd and renew'd as oft as they die," he wrote. The spirits were attacked first by the plague, Willis claimed, and so they had to be fortified. "Therefore wine and confidence are a good preservative against the plague."

The plague did not come to Willis in Oxford, but fugitives from London soon did, including Charles II and his court. Twenty-three years earlier, the king's father had come to Oxford and left it a ruin. Now Charles II arrived, stuffing the city with his own retinue. Oxford became so cramped that sixty lawyers who had business before the courts slept in a single barn.

Dedicated royalists such as Willis could feel proud to host their new king, but they could also see his flaws up close. Charles was now wooing a new mistress, a young virgin named Frances Steward, but he had not abandoned Barbara Palmer. According to one rumor, he caught Barbara in Oxford with another lover, whom he simply waved away with the command, "Go, you rascal." The king showed more concern for Barbara's second pregnancy than with the march of the plague. Oxford might be free of disease, one observer noted, but the king had introduced "the infection of love."

Parliament followed the king to Oxford. The only attention it gave to the plague was to pass laws to protect wealthy people from house inspectors. Dissident ministers were heroically helping victims of the plague, but Parliament stepped up its persecution of them by banning them from coming within five miles of their old churches. Charles's court refused to talk about the epidemic at all, amusing itself with the latest gossip about the king and his mistresses. The irascible Anthony Wood looked on the visitors with scorn. "They, though they were neat and gay in their apparel, yet they were very nasty and beastly, leaving at their departure their excrements in every corner, in chimneys, studies, coal houses, cellars. Rude, rough whoremongers; vain, empty, careless."

Willis was called from time to time to tend to the royal family for the occasional illness or miscarriage. The poor foot soldier who had defended Charles's father now entered the king's court as a wealthy, revered physician. But Willis never became an official physician to the king. For all his success, he was still a blunt, plain-speaking, small-town doctor with none of the cosmopolitan grace required for a life at court. The duchess of York, the wife of Charles's brother James, asked him why she suffered a series of miscarriages. To which Willis answered in Latin, "*Mala stamina vitae*"—there was an unlucky thread woven by the Fates into her life. The implication was that syphilis or some hereditary weakness in the royal family was to blame. When the king heard about the remark, he banned Willis from the court for the rest of his life. Later, he groused that Willis had rid him of more of his subjects than any enemy army.

One consolation that the plague brought Willis was the chance to see old friends from the Oxford circle again. John Wilkins, William Petty, and other virtuosi left London for Oxford to wait out the epidemic, passing the time with talk and experiment. Richard Lower showed them some of the experiments he had been trying out with Wren's injections. Even as Lower had been working with Willis on the brain, he had wondered if injections would keep a dog alive instead of food. He had injected a quart of warm milk into a dog, but the animal died in an hour, its blood curdled. Lower then wondered if he might have more success if he fed a dog with another dog's blood. Perhaps the same could be done to people; even a sheep's blood might be able to save a patient from bleeding to death. The blood would have to flow directly from a donor, Lower realized, or it would clot. Boyle helped Lower in his first attempts. They opened up the jugular vein of a dog and attached a pipe to it, which they then attached to the jugular of a second dog. The experiment failed, the blood clotted in the pipe, and the dogs died. Boyle was eager to try some new variations on the experiment, but Lower left Oxford in August 1665 to go to his home in Cornwall to search for a wife. Aeson's revival would have to wait.

As Willis watched his assistant struggle with transfusion, he was still turning over their work together on the brain. How could the things they had discovered help him understand the afflictions he had seen in his patients, especially convulsions? During the epidemics of the 1650s, he had observed the awful thrashings that meningitis could cause. Even as he had started his dissections of the brain with Lower, he had been confronted by fresh examples in an outbreak of a mysterious new kind of fever that robbed its victims of speech and knowledge, leaving only deliriums and nightmares. Willis treated one feverish boy who "talked idly, complained that his cap was fallen into the water, and by and by becoming speechless, within four hours, whilst I was sent for, he expir'd before I came."

A few days later, the boy's younger sister fell ill with the same disease and began to babble that her coat had fallen into the water. She was gripped by convulsions, which lasted a long, awful day, and then she died as well. For Willis, dying children were a horribly common sight. Of his own eight children, four had died by the mid-1660s. Watching this girl follow her brother to death, he was filled with frustrated rage. Unable to cure them, he did the next best thing: he got permission from their parents to perform an autopsy of the girl, "to find out from her death," he wrote, "the knowledge of the aforesaid disease."

Convulsions, Willis knew, did not always kill their victims. During the year of the plague, a woman Willis later described as "an illustrious virgin" became terrified of dying. The terror sent her into convulsions twice a day every day, taking over her body promptly at eleven o'clock and five. Willis came one morning to see a morning's convulsion for himself. At ten she was herself, "so that none would ever suspect her to be sick; at eleven of the clock she began to complain of a fullness of her head, and numbness of spirits, with a light swimming." A convulsion came on, and her servants put her in bed and sat on her with pillows.

Willis began to sketch out new theories about epilepsy and other kinds of convulsions. In the seventeenth century, Europeans still interpreted convulsive diseases with a mixture of Galen,

Christian theology, and ancient traditions of magic. Writing about hysterical fits, for example, Willis noted that that "most ancient, and indeed Modern Physicians, refer them to the ascent of the womb, and the vapours elevated from it." In the 1620s, Harvey had claimed that hysteria came "from unnatural states of the uterus." Willis himself had thought the same thing about hysteria in 1650, when he was still a green doctor, but he had grown dissatisfied in the years since. Doctors were too quick to blame any unusual symptom in a woman on hysteria, as if the womb were the source of all their diseases. Once, after a woman who had been troubled with "the mother fits" died, Willis performed an autopsy and found the womb "wholly faultless." In his experience, hysteria was not limited to fertile women, who were supposed to be its only victims. He saw hysteria "in maids before ripe age, also in old women after their flowers have left them; yea, sometimes the same kind of passions infest men." When Willis tied off the nerves that innervated a dog's chest, he could produce the rapid pulse, the shortness of breath, and all the other symptoms "imitating the type of hysterical attack."

Epilepsy had its own confused history. In ancient Greece epilepsy was thought to occur when phlegm flowed out of the brain, cooling the blood and cutting off the flow of air from the lungs. The breath trapped in the body began to foam and fizz, making the arms and legs flail. If the attacks were too severe, Hippocrates warned, parts of the brain might melt to water.

Galen, as usual, had a more elaborate explanation. He claimed that either black bile or phlegm could cause epilepsy by rising as vapors into the brain, where they condensed into fluids and clogged the ventricles. There they blocked the spirits from flowing into the nerves, which responded by shaking violently.

At the same time, though, epilepsy had gained a reputation as a sacred disease. Babylonians had considered it the work of a demon or a departed spirit. An epileptic fit was a bad omen to the Romans, and magicians warded off fits with spells or drinks of human blood. Europeans blended these traditions with the Bible's teachings. The church taught that evil spirits could cause epilepsy, pointing to the way Jesus had cured an epileptic by driving the spirits from his body. The idea that epileptics were possessed made sense: an epileptic seizure looked as if its victim was literally seized by a spirit and tossed around, only to be set free and returned to his former self.

Willis would never deny that the Devil could cause people harm, but the Prince of Darkness didn't have to break the laws of nature to do so. "He is not able to draw more cruel arrows, from any other quiver, or to show miracles by any better witch, than by the assaults of this monstrous disease," he wrote. Willis rejected old explanations of convulsions, looking for a new one based on the chemistry that the brain used to control the body. He came up with what he called a "clean, new, and unusual hypothesis." A doctor should look for the source of all convulsions not in the ventricles, in the womb, or in the supernatural world. He should look in the brain.

Willis saw the arteries feeding the brain as an alembic designed for distilling spirits from the blood, leaving the bigger, more sluggish particles behind. Spirits flowed peacefully from the brain into the nerves, and when they reached the end they encountered sulfurous particles in the blood, setting off an explosion. The body's explosive power was normally unleashed only at the right time in the right muscles. If sulfur was somehow planted in the brain itself, however—perhaps by the Devil or by some natural happenstance—it might mix with the spirits there. Spirits driven from their normal paths by passions might be particularly vulnerable. The spirits and sulfur might explode, setting up a chain of explosions all the way down through the nerves "like a long train of gunpowder."

Willis traced the course of hysteria, epilepsy, and other kinds of convulsions from the foreshadowing in the head to the thrashing of the limbs. He brought an unprecedented precision to convulsions, just as he had to the brain's anatomy. Looking over his description of epileptic seizures, one modern neurologist has written, "This is precisely the modern view of the nature of constitutional epilepsy, if we substitute the idea of an electric discharge for a discharge of animal spirits."

The deliriums and depressions and ravings of hysteria likewise became for Willis a matter of explosive chemistry. It did not occur to him that the psychological life of his patients might have the power to produce hysteria—nor did he think that it could offer a cure. Willis may have made a great advance by dismissing the womb, but he still prescribed the traditional treatment for hysteria: tying a bandage around the bellybutton. Older doctors claimed the bandage stopped the womb from wandering, but Willis—as usual—had a new explanation: it tamed the racing animal spirits. (For once, at least, Willis also prescribed a remedy that wasn't painful, foul, or harmful. He recommended that hysterical women enjoy "the Pleasures of Venus.")

Willis set down his thoughts on convulsions in a new book he entitled *Cerebral Pathology*. Lower wrote eagerly to Boyle that "there is not a disease of the head which he does not excellently illustrate with very rare observations and cases." As Willis was finishing the manuscript in the spring of 1666, the plague finally began to ebb from England. In London, the weekly death rate dropped from thousands to hundreds, from hundreds to dozens. Its final toll was devastating. Out of 400,000 Londoners, almost 100,000 were dead, and in the countryside crops rotted unharvested. England slowly recovered through 1666, but in September a fresh disaster hit. London went through a fiery convulsion of its own, an explosive fit far more destructive than any epileptic seizure.

One Sunday morning, a bakery oven in Pudding Lane burst into flames. Strong winds carried the fire across the neighborhood, detonating a chain of warehouses full of oil, tallow, hemp, and spirits. Paving stones grew so hot that they exploded like grenades, while molten lead flowed in the gutters. The fire became so big that sixty miles away in Oxford, Anthony Wood recorded that "the sunshine was much darkened. The same night also the moon was darkened by clouds of smoke and looked reddish. The fire or flame made a noise like the waves of the sea."

As the fire spread, people rushed to the Thames to escape in boats. Houses were torn down in the hope of stopping the fire from spreading farther, but it kept growing. It raged after the sun set, making the night "as light as day for ten miles roundabout," according to the diarist John Evelyn. The fire burned through Monday and reached its peak on Tuesday, even pulling down St. Paul's Cathedral. One observer wrote, "I believe there was never any such desolation by fire since the destruction of Jerusalem, nor will be till the vast and general conflagration." Only on Wednesday did the fires finally die, having claimed thirteen thousand buildings over 436 acres. "London was, but is no more!" Evelyn mourned.

But London did have a future, and Thomas Willis would be a part of it. In the wake of the plague and the fire, Willis would come to London and find his greatest fortune. More important, he would finish what he had started with *The Anatomy of the Brain and Nerves*. He would now map the anatomy of the soul.

Chapter Ten

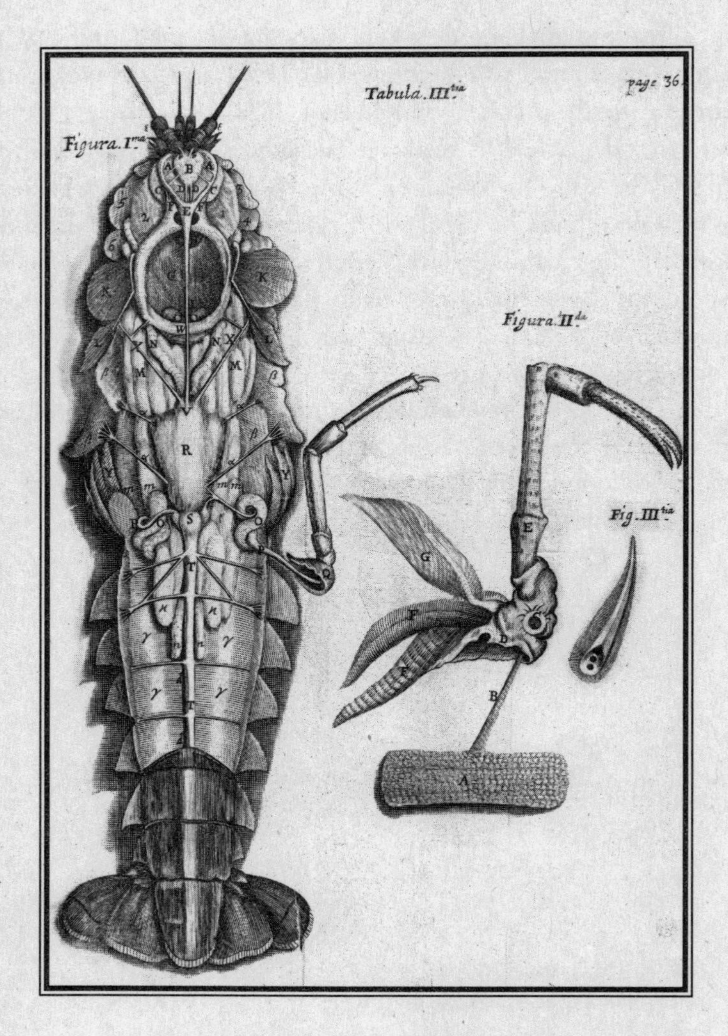

THE NERVOUS SYSTEM OF A DISSECTED LOBSTER, FROM WILLIS'S *TWO DISCOURSES CONCERNING THE SOUL OF BRUTES.*

The Science of Brutes

A friend of Lord Conway sent him a letter from London a few weeks after the fire had died. "Men begin now everywhere to recover their spirits again," he wrote, "and think of repairing the old and rebuilding a new City."

As London slowly recuperated from plague and fire, the Oxford circle shifted for good to the capital. Christopher Wren had been traveling through France, absorbing its architecture, when the plague hit. He returned to London, and a few days after the fires died away he presented King Charles with a plan to rebuild the city from scratch, with magnificent boulevards, piazzas, and a new cathedral. His map was a blueprint of the Royal Society's ambitions, a plan for creating a powerful, efficient England based on reason rather than tradition. Not even Charles II could accommodate the scope of Wren's dreams. He asked Wren instead to design a new St. Paul's Cathedral and dozens of other buildings. Robert Boyle packed up his chemistry equipment in Oxford and came to the city to help run the Royal Society. Even Willis himself, the only native Oxonian of the circle, felt London's pull when Archbishop Sheldon asked him to come to London to help rebuild its medical ranks. For Willis, a request from an archbishop was as good as a command, and he moved his family to a house in St. Martin's Lane.

By the time Willis arrived, the Oxford circle was dazzling London. Christopher Wren was joined in designing new buildings by Robert Hooke, and together the two old friends brought a geometrical elegance to London unlike anything it had seen before. Robert Boyle was as sick as ever (he had taken to blowing dung powder into his eyes to stop his sight from failing), but his air pump had made him the most famous natural philosopher in England. With the publication of a string of books in which he described his research at Oxford, he won a place for chemistry at the high table of science and laid out the experimental method.

The most sensational member of the circle proved to be Richard Lower. After finding a wife in Cornwall, he had returned to Oxford in 1666, ready to take up his work on blood transfusion again. His previous experiments had failed because the blood from one dog clotted before it could get into another. So he now tried a new method, inserting one end of a pipe into an artery of a donor dog and the other into a vein of the recipient. Following Harvey, he was convinced that the blood flowed more strongly in the arteries than in the veins and so it would push its way out of one dog and into the other. The new arrangement did the trick. Lower bled a dog nearly to death and then transfused it with a fresh supply. It leaped down off the table, licked Lower happily, and then rolled on the grass to clean off its fur.

The news of Lower's success moved the Royal Society to run some experiments of its own. They poured the blood of calves into sheep and of lambs into foxes. The sheep survived, the foxes didn't. A few months after Willis moved to London in 1667, Lower followed him to the city, setting up a medical practice of his own. He also took over the Royal Society's transfusion work and set out to try something truly spectacular: transfusing blood into a man.

Finding that man was not easy. Bedlam, London's lunatic asylum, refused to surrender a patient for such a wild-eyed experiment. Finally Lower found his subject in the church led by John Wilkins, a Cambridge-educated tramp named Arthur Coga. Lower later wrote that Coga "was the subject of a harmless form of insanity." Having explored the connections from the blood to the brain so carefully a few years earlier, Lower was convinced that transfusions could "improve his mental condition."

With silver pipes and quills, Lower linked a vein in Coga's arm to the carotid artery in the neck of a lamb. (Coga liked to say later that the lamb's blood was special because Christ was the Lamb of God.) Lower carried out transfusions twice on Coga, and his subject survived both ordeals. Afterward he enjoyed a pipe of tobacco and a glass of Canary wine in front of forty witnesses. All London wondered now what could be done with Lower's invention. Maybe the old could be rejuvenated with the blood of the young. Could madmen be healed with sober blood? Could Quaker blood infect a man with its fanaticism?

The sensation did not last very long. Coga quickly embarrassed his experimenters, raving instead of praising, and using the guinea Lower gave him to get drunk. "The wildness of his mind remains unchanged," wrote Henry Oldenburg. Soon afterward, French experimenters carried out a transfusion of their own that ended in disaster, killing their volunteer. The Royal Society abandoned transfusions, although Lower remained convinced that the procedure could be useful. Ten generations would pass before anyone would dare pour blood into a human again.

Lower still had plenty of other experiments to do. He set out to meet one of the questions William Harvey had raised but not answered: how does the heart pump blood? While the circulation of the blood was universally accepted by the 1660s, some physicians claimed that it was not the heart itself that drove the blood in a circle. Within the heart, they argued, some ferment heated up the blood and made it expand, driving it into the arteries. Thomas Willis held a middle view, arguing that the heart was a pumping muscle and that it also housed a ferment that made the blood warm and red. After Lower moved to London, he slowly realized that his old teacher had made a profound mistake.

Lower's doubts emerged from his work with Willis on the brain. "If the blood moves through its own power, why does the heart need to be so fibrous and so well supplied with Nerves?" he asked. Lower decided to look carefully at the heart's muscle fibers. He discovered that they were wound in spirals, so that when they contracted they compressed the heart's chambers. Harvey had simply claimed that the heart was a muscle that drove blood out of its chambers—but Lower showed how. He also showed that in the heart's violent contractions, everything in the chambers was squeezed out along with the blood. No warming ferment could linger for long in the heart, and once it was expelled, it could not get back in. Lower injected dye into the blood vessels of the heart itself. He found they could not find a way into the heart's interior. The heart, Lower realized, did not even need blood inside its chambers to beat. Even when he replaced much of a dog's blood with beer and wine, its heart continued to throb. The heart, Lower declared, was a muscle, pure and simple, acting as a piston. The blood did not get warmed in the heart; the blood warmed the heart and the rest of the body.

Lower's work raised the question of whether the heart was the place where blood turned red. Willis had been impressed that a bowl of blood formed a red top layer, but he had never bothered to do the simplest of experiments: skimming off that top red layer. He would have discovered, as others did in the mid-1660s, that once the darker blood below was exposed to the air, it turned red as well. The color did not come from some particular group of particles in the blood but from something in the air. Could this be the same substance that Boyle pumped out of his glass chamber, the substance on which a lark's life depended?

Robert Hooke, who had built Boyle's pump, had already begun to look into this idea with one of the grisliest experiments ever carried out by the Oxford circle. He based it on the fact that even the slightest nick to the membrane surrounding a dog's lungs could stop it from breathing. In 1664, he rigged a set of bellows with a long piece of cane at its tip. He cut open the windpipe of a dog so that he could slip the cane into its airway. While he kept the bellows pumping air into the dog's lungs, he cut open its chest to see what was happening inside. As long as he pumped, the lungs swelled and shrank and the heart beat steadily. Hooke kept the dog alive by pumping bellows for an hour. When he finally stopped, the lungs went slack and the dog's heart went into convulsions.

The results were tantalizing, but Hooke was disgusted by the experiment. The Royal Society urged him to carry it out again, but he refused "because of the torture of the creature."

It took three years for Hooke to return to the project, and only when Lower came to London and offered his dissecting skills. This time Hooke and Lower nicked the dog's lungs so that the air they pumped into it would leak out. Then they added a second pair of bellows to the experiment, so that they could pump a continuous stream of air into the dog. Now the dog could breathe without moving its lungs at all. Lower and Hooke found that they could keep the dog alive as long as they pumped. When Lower sliced a piece of the dog's lung away, he could see that blood was continuing to flow through the lung's fine vessels. The new round of experiments showed that it was not the mechanical flow of air into the lungs that was the secret to life but the air's contact with the blood.

Hooke apparently got over his qualms about torturing dogs. He now fitted a bladder over one end of a brass pipe and fitted the other end into a dog's exposed windpipe. The dog had to breathe the same air over and over again, and within a few minutes it struggled violently. In eight minutes it was near death, but Hooke could revive it by letting the dog breathe fresh air again. It was becoming clearer with each investigation that something in the air was consumed during breathing and chemically transformed into another substance that was then exhaled. Whatever it was that breathing brought into the body turned the blood red. Lower opened the chest of a dog and slit its windpipe. He then corked its exposed end so that the dog could not get any fresh air. He slit open the dog's carotid artery and found that the blood flowing through it had become as dark as the blood in any vein. But the darkness of the blood was not permanent. If Lower poured it into a dish, it turned bright red.

The final proof Lower needed came when he set up Hooke's double-bellows experiment again. It allowed him for the first time to cut into the pulmonary vein that traveled from the lungs to the heart, even as a dog was still breathing. If the theories of Willis and other physicians were correct—if blood turned red only in the right side of the heart before it entered the arteries—Lower predicted that the blood in the vein should be dark. But Lower discovered that even in the pulmonary vein, it had already turned a bright crimson. The blood surging through the lungs had been exposed to air as if it were sitting in a dish and had been turned red.

There was one more step in Lower's series of discoveries to be made, but he did not take it. For reasons lost to history, Lower stopped his extraordinary run of experiments in 1670, went back to his busy medical practice, and never wrote another scientific work. (He even stopped paying his dues to the Royal Society and was expelled in 1675.)

The final step belonged to a young Oxford graduate named John Mayow. Mayow had listened to Willis's lectures and paid close attention to the experiments of Hooke and Lower. He then proposed that all their experiments made sense only if the purpose of breathing was to absorb particles that turned the blood red and were carried by the blood to the muscles. In tiny explosions, the muscles used up the particles and had to be replenished with more.

The heart, as Lower had shown, was a muscle like any other, and it needed these particles as well. That was why the heart beat more quickly when the body worked hard—to supply the body with more particles from the lungs. That was also why a lark died in an evacuated chamber—because its heart ran out of these explosive particles and stopped beating. Mayow tested his hypothesis by opening up a dog and inserting a tube into its vena cava, the major vein carrying blood to the heart. He waited until the heart and lungs had stopped moving and then blew air into the heart. The heart leapt back to life. Particles, Mayow argued, were responsible not only for the beating of the heart but also for making fires burn and for fermentations of the blood.

Willis was proven wrong, but he had taught his students well enough for them to be able to move closer to the underlying reality. He accepted some of their conclusions, now calling the heart "a mere muscle consisting of only flesh and tendon." This was no small point on which Willis's students were correcting him. Ever since Plato, the lungs and the heart had been considered the center of the vital soul. It was from the vital spirits that animal spirits were created. In 1659, Willis had taken the first steps to a new definition of those spirits. Now, ten years later, Lower, Hooke, and Mayow were finishing the job. They never managed to discover oxygen, the substance that all living things must breathe, but as they reached, their fingertips brushed across the truth.

—

Willis did not see Lower make a dog breathe. After his move to London, Willis rarely went to the meetings of the Royal Society, because he was too busy building what would become the first major medical practice in England's history. Archbishop Sheldon introduced him into the highest circles of London society, and noblemen, rich merchants, and leaders of the church became his patients. The boy who had fought for table scraps as a servant was now buying country manors with twenty hearths apiece. But Willis remained modest and devout. He declined a knighthood, it was said, because he was not "desirous of a distinction by any dignity or rank in his own person." He began each morning with prayers at six and then spent several hours treating poor patients without charge at his house. He set out in his coach to visit wealthy patients around the city. Along the way, he might stop off at his apothecary or get the latest gossip at his favorite coffeehouse. In the afternoon, Willis might supervise a surgical operation or ride into the country to see more patients. At five he was at prayers again, and after dinner he made notes on his cases and went on writing into the night, giving rein to his speculations but always ready to answer a late call for help. All of the fees he collected on Sunday, his busiest day of the week, he gave to charity.

With his move to London, Willis became the most famous physician in all England, perhaps in all Europe. "He became so noted," Anthony Wood wrote, "and so infinitely resorted to, for his practice, that never any physician before went beyond him, or got more money yearly than he." His theories about how people became sick and how the body worked were hugely popular. Yet his remedies remained, as ever, generally useless. He continued prescribing his eclectic blend of medicines—a blistering plaster for one patient, a ground-up millipede for another—and claimed success when his patients recovered and escaped blame when they didn't. If they recuperated, they usually did so in spite of his attentions. His track record was no better for disorders of the brain. For epilepsy, he gave his patients emetics to make them vomit, bled them with leeches, fed them peony roots and wolf livers, and had them wear amulets filled with mistletoe.

Yet Willis also became the first truly great neuroscientist, thanks in part to his medical reputation. He could persuade people to let him dismantle their dead husbands, wives, and children, and even dissected the bodies of aristocratic patients. For the first time in the history of medicine, Willis could link the diseases and disorders that people experienced in life to the abnormalities he found in the brains after death. If he had been cutting open the brains of criminals, he could not have made the link so persuasively. His readers might have thought that the pathology he saw was just a sign of the low class of the people from whom the brains came. Because the brains belonged to England's ruling class, it became hard for his readers to dismiss his observations. The respectability of his success allowed Willis to expand his mechanical, chemical explanations of the brain to include the soul itself without being accused of heresy.

Descartes came to believe that the soul was a thinking substance altogether different from the matter that made up the universe. Animals were automata without souls. Willis had found that animal brains, particularly those of mammals, had some astonishing similarities to those of humans, which were reflected in the way they lived. They could perceive, be aware, even learn. As a boy, Willis had watched horses learn the routes to meadows and ponds; he had watched dogs teach other dogs to hunt. He wrote about "the most admirable republics of bees and ants, in which, without any written laws or promulgated right, the most perfect ways of government are exercised." Like humans, animals could acquire knowledge—what Willis called "the science of brutes."

And yet Willis believed that animals do all this despite being made only of matter. He took a stand against Descartes and his followers, who "endeavouring as much as they could to discriminate the soul of beasts from the humane, affirmed them to be not only corporeal and divisible, but also merely passive."

Willis was convinced that matter could be active. A body could think. Willis drew his inspiration from Pierre Gassendi, who had already served as a crucial influence on the Oxford circle by offering a philosophy of atoms that could be squared with Christianity. Gassendi had also claimed that animals had material souls capable of some of the things Descartes reserved for the human soul. But Gassendi was a priest, not an anatomist, and he believed that the sensitive soul "is infinitely above the reach of our senses."

Willis, on the other hand, was confident he could sketch out the anatomy of this soul. Animals had mortal, material souls that encompassed their nerves and brains. Man was a partial exception, Willis argued, a "double-soul'd animal," blessed with both a material soul and an immaterial one.

To understand the material soul, Willis dissected more brains. For his new assistant, Willis turned to Edmund King, a gifted surgeon who had helped Lower pour blood into Arthur Coga. Willis and King looked even more closely at the human brain, using a penknife to scrape away the gray matter, leaving behind "chords or strings, as it were distinct Nerves, wonderfully communicating among themselves." Willis dissected the brain of a monkey for the first time. He even looked at animals without backbones. Willis and King dismantled a zoo of fish, lobsters, oysters, and earthworms.

Willis concluded that vertebrates and invertebrates alike stay alive in the same fundamental way, by absorbing particles into their blood. Willis discovered earthworms had passageways through their skin that let them breathe without lungs; microscopic pores in the shells of insects let them do the same. He injected ink into the gills of fish and found that it traveled into the animal's heart and then into its arteries. He speculated that it must draw something out of the water much like what we draw out of the air when we breathe. Pierre Gassendi had described the sensitive soul as a flame, and Willis, with his knowledge of chemistry and anatomy, agreed. It was a fire that burned without light throughout the creature, and the particles that it took in from air or water were the fire's fuel. Death came when the soul's flame was extinguished.

Just as all animals shared a vital flame, they all had nervous systems to control their bodies. Willis cracked open the shells of lobsters and followed the branches of their nerves; he teased apart the mesh of nerves in oysters. Earthworms had clumps of nerves, "whitish like a bubble," he wrote, that he decided must be tiny brains. He was looking at the other part of the material soul that Gassendi had proposed, which sensed the outside world and made the animal breathe and move. Of course, worms and dogs didn't behave the same way, and that was because they had different kinds of souls—and therefore different kinds of nervous systems. They were arrayed on a ladder from lower to higher, designed for their particular station.

In both animals and man, the sensitive soul consisted of spirits that flowed up and down the nervous system and wandered the flesh of the brain. The chords and strings of the brain were like the invisible corridors of the soul's palace, through which its spirits traveled. Willis compared the structures within the brain to spacious courts, porticoes, and walks. He called the spine "the King's highway," on which spirits moved like soldiers between the soul's palace and its bodily kingdom. Willis was impressed by the complex designs of the brains of dogs and cows. Monkeys' brains, he discovered, were even more like humans' than any brains he had seen before. Like humans, monkeys had huge brains for their body size, and several parts of their brains "came nearer the figure and magnitude of those parts in a man."

Depending on an animal's brain, Willis believed its soul might be endowed with memory and imagination and other capacities of human minds, which were carried out in the same parts of its brain. An animal's senses conveyed light and sounds and smells through its brain "like a waving of waters," as Willis described it. This waving traveled into the animal's cortex, where the brain perceived the impressions and sometimes imprinted them as a memory. Later, if the animal experienced the same set of sensations, the spirits would travel along the same pathways, making its body move the way it had moved the first time.

A horse, for example, grazing barren earth, would feel hunger as its stomach stirred up the animal spirits in the surrounding nerves. "The spirits being thus moved by accident, because they run into the footsteps formerly made, they call to mind the former more plentiful pasture fed on by the horse, and the meadows at a great distance," Willis speculated. The horse's memory of the pastures would then trigger its memories of walking to them, which in turn would send its spirits out to its legs to bring the horse there.

Memories, habits, perceptions, imagination—the human brain created all these in the same way as the brain of a horse. Willis believed that the sensitive soul was responsible for many other capacities of our minds, including wit, ingenuity, and affection. For Willis, emotions were produced not by the four humors or the heart but by the movements of this soul. Love was an attraction of spirits toward a thing or its mental image. Anger was "like the sea working with opposite winds and floods excited from every coast." When we feel happy, our spirits expand, spreading out from the brain and making our body feel elated. Sadness withdraws the spirits deep within the brain.

According to Willis, the spirits could not flow through the nerves and the brain without a rest. They needed a pause from time to time on their journeys. The spirits of the cerebrum needed to settle down for a few hours every day, withdrawing into the brain and becoming restrained "as it were with chains," Willis wrote, "that they may not enter into motion." With the animal spirits of the cerebrum locked within the brain, a sleeper could not move. But a sleeper still breathed and his heart still beat, because the spirits of the cerebellum controlled these organs. Because they had only simple work, they needed only momentary rests.

In this light, sleep and its disorders became less mysterious. Willis had encountered narcoleptics who fell asleep suddenly, even with their mouth full of food. Although many considered the condition to be some kind of evil habit, Willis recognized it as a disease, speculating that it was caused by too much blood flowing into the vessels of the brain, cramping the animal spirits and making them unable to circulate. For narcolepsy and other forms of sleepiness, he prescribed bleedings, a purge, and then medicine to hold back the "watery deluge" as he called it—including a drink made from the miraculous new bean, coffee.

With sleep came dreams, and here too Willis sought to ban the supernatural. Dreams were often seen as divine visions or visits from demons. Alchemists sometimes claimed that they received their recipes directly from God in their sleep. Nightmares were supposedly created by an evil incubus crouching on the chest. Willis argued that dreams were the result of a few spirits slipping out of their restraints during sleep and wandering through the brain, causing mischief as they went, "without any guide or ruler." Sleepwalkers, on the other hand, wandered and even spoke without being aware of what they were doing. The wandering spirits within their brains, Willis argued, had stirred up others that traveled down the spine and caused them to walk. Since most of their spirits were still locked down deep in the brain, they couldn't form any memories of the experience.

To the extent that animal and human brains were alike—and they were alike in many ways—Willis believed that animal souls and human souls must also be alike. Whatever differences he could see must represent differences between them. "The smelling Nerves," as Willis called them, are bigger in animals than in humans—"the reason of which is," he wrote, "because they discern things only by the sense, and especially their food by the smell; but Man learns many things by education or nurture and discourse, and is rather led by the taste and sight, than by the smell in choosing his aliments." Man, on the other hand, has a much bigger cerebrum and a far more furrowed cortex than any animal. That meant memories must be housed there, Willis argued, because humans obviously have a much larger warehouse of memories than animals.

Yet Willis also believed that humans alone had a rational soul, and that his own study had confirmed that it was immaterial. On the one hand, humans had to have a rational soul, because they could obviously think of things far beyond the scope of an animal brain. If that rational soul was material, Willis would have expected to find a corresponding difference between the brains of animals and humans, but, Willis wrote, "we have noted little or no difference in the head of either." The astonishing similarity of human and animal brains meant that the rational soul could not be a material thing.

Once a baby had developed far enough in the womb, Willis believed, God created its rational soul, which He placed in its brain. The rational soul had no substance and therefore knew no death. Even after the vital flame of the body was snuffed out and the sensitive soul died, the rational soul survived. To that extent, Willis agreed with Descartes, but only that far. Ultimately, Willis was an anatomical Gassendi. He put the rational soul in the corpus callosum, the white slab of brain onto which the spirits of the senses projected their perceptions. Willis's rational soul could oversee the sensitive soul, which could handle most of the work of living on its own. Just as a king shouldn't bother with the details of clerks, it was not fit for the rational soul to busy itself with these "lower offices." Instead it viewed the perceptions formed by the sensitive soul like a king watching a royal masque. "The rational soul, as it were presiding," he wrote, "beholds the images and impressions presented by the sensitive soul, as in a looking glass, and according to the conceptions and notions drawn from these, exercises the acts of reason, judgment, and will." The rational soul could draw abstractions and raise its thoughts to things far above what the senses could offer—"God, Angels, It self, Infinity, Eternity."

The rational soul was, in other words, the king of the body, sitting on his throne at the center of the palace of the brain. But power, as Willis had seen for himself, came with vulnerability and constraints. The rational soul could not have any direct knowledge of the world. It could only gradually reach understanding by reasoning about what the sensitive soul presented it. In a healthy brain, the sensitive soul peacefully operated the body, its spirits moving in and out of the brain in harmony, accurately describing the outside world to the rational soul and accepting the rational soul's guidance. Some diseases, such as Lady Conway's migraines, disturbed only the edges of the nervous system. Describing her case, Willis wrote that the migraine, "having pitched its tents near the confines of the Brain, had so long besieged its regal tower, yet it had not taken it; for the sick Lady . . . found the chief faculties of her soul sound enough."

But the brain itself was an organ, and that meant that its souls could fall prey to illnesses. Willis called these diseases of the soul "Civil Wars."

"The lower soul," he wrote, "growing weary of the yoke of

the other, if occasion serves, frees itself from its bonds." The two souls then vie for power like two armies planted within the brain "till this or that champion becoming superior leads the other away, clearly captive."

Unwilling to submit to its ruler, the sensitive soul might become overwhelmed with melancholy. Although Willis had abandoned Galen's old system of humors, he didn't dispute that melancholy was real. It was the underlying explanation that Willis challenged. "We cannot here yield to what some Physicians affirm, that Melancholy doth arise from a Melancholick humour," he wrote.

Willis made the nervous system a chemical reactor, with the spirits like rays of light produced by the blood's flame. A lamp burning wine or oil or fat produces different kinds of light, and likewise the body's animal spirits changed if the soul's fuels changed. If the spleen stopped filtering the dregs from the blood, it could become too salty, turning the transparent animal spirits dark and sooty. Instead of the darkness of black bile, Willis offered the darkness of the spirits themselves. As they traveled the nerves to the brain, the spirits presented the rational soul with a dark, distorted image of the outside world, which the soul took as reality. And because the spirits were tainted by the salty blood, they corroded their way through the brain, carving new, contorted paths that permanently distorted a person's thought, turning the familiar to the strange, the bright to gloomy.

If a brain became diseased enough, Willis believed that the rational soul itself might become permanently affected. In delirium, the rational soul was presented not with realistic perceptions but with wild fantasies. Frenzies turned the peaceful streams of animal spirits into a "storm of waters raging in a tempest." They could spill over their banks and blaze new paths through the brain, driving a person into permanent delusions. If the spirits took on a corrosive chemistry and became uncontrollable, their victim went mad.

Willis's advice for melancholy included some very ordinary

suggestions. "Pleasant talk, or jesting, singing, music, pictures, dancing, hunting, fishing, and other pleasant exercises are to be used," he wrote. But to him, these pastimes were ultimately mechanical cures: they pulled the spirits back onto their proper paths through the brain. When he faced violent madness, Willis abandoned the fishing or dancing and called for harsh treatments to bring wild spirits into submission. "Furious Mad-men are sooner, and more certainly cured by punishments and hard usage in a strait room, than by Physick or Medicines."

Willis did not cast all mental disorders as civil wars between the two souls of man, however. Some of them passed above the sensitive soul and, in his words, "most chiefly belong to the Rational Soul"—despite the fact that they depended "upon the faults of the Animal Spirits, and the Brain it self." Willis called these conditions "stupidity" and "foolishness."

What Willis called stupidity roughly corresponds to modern diagnoses such as Alzheimer's disease and mental retardation, while his foolishness is more akin to psychosis or schizophrenia. Psychiatrists today can recognize some of their modern conceptions of these conditions in Willis's writing. Go any further back in time, though, and the trail gets muddy. In ancient Greece, for example, physicians and philosophers had nothing to say about people with defective brains who were permanently incapable of abstract thought or sound judgment. Such categories simply didn't exist. The Greeks did speak of *phronesis,* the power of the soul to perceive things, which Hippocrates said was created by "the proper blend of the moistest fire and the driest water." Throw this blend out of balance, he warned, and *phronesis* would suffer. If the water mastered fire, the soul would slow down, unable to meet the quick perceptions of its senses. Too much fire, and the soul would move too quickly, rushing to judgments and becoming confused. But to the Greeks, a person with too much fire or too much water was not locked in some genetic prison; these were just snapshots from the soul's never-ending flux.

Even in medieval Europe, these categories didn't exist in the medical and psychological dimensions in which they are found today. Fools were not shunned; in fact, they were often treated instead as the blessed children of God. The word "idiot" did not yet apply to a discrete group of intellectually handicapped people. The courts recognized people they called *idiotae,* but this was a legal term for people who had their estates taken away because they could not pass a test, such as counting to twenty. Neither idiots nor fools had yet become biological creatures.

Thomas Willis created some of the earliest medical definitions for them. He kept notes on stupidity that ran in families, of the way fevers could cause or even cure foolishness. He sometimes dissected a patient whom he referred to as "slow" or "a fool from his birth," looking for a physical cause for his condition. The brain of one foolish boy seemed wasted and shrunken. Willis was so struck by it that Christopher Wren made an engraving of the brain, which appeared in *The Anatomy of the Brain and Nerves.*

To explain stupidity and foolishness, Willis poured the old Hippocratic wine into the new bottles of the scientific revolution. Both were disorders of animal spirits "leaping forth, or running out desultorily or after a leaping manner." But Willis's dissections of brains also convinced him that a misshapen brain could also be the cause. A small brain might not generate enough spirits to make the mind sharp. A deformed head might make the animal spirits bounce around the brain at the wrong angles. The brain itself might be too thick for the spirits to cut their tracts—a condition Willis claimed was common among "those that are born of Plowmen and Rusticks, as if they were formed of a worser clay."

To call a peasant an idiot was quite ordinary for the intellectual elite in Willis's day. In the Renaissance, all people except philosophers and theologians were called *homines idiotae,* because they couldn't rise from particulars to universal truths. By the late 1600s, the ranks of the educated and literate were swelling in England, and only servants and peasants remained idiots. Willis simply added a medical tone to this distinction. But he did not believe that stupidity was an all-or-nothing matter. It came in degrees, "for some are accounted unfit or incapable, as to all things, and others as to some things only." Some were stupid in "the learning of letters" but good enough for mechanical arts. Others who couldn't handle either might be good at farming—Willis's Plowmen and Rusticks. Others couldn't even farm, able only to learn "what belongs to eating or the common means of living." And below them were "dolts or driveling fools," who "scarce understand any thing at all, or do any thing knowingly." The scale that ranked worms, cows, and men also ranked the men against one another.

People could be born stupid or become stupid with age. For Willis, senility meant that animal spirits fermented until they spoiled like good wine gone bad. He recognized that long-term epilepsy could bring on stupidity. Sometimes stupidity came before paralysis. Fevers could bring a form of stupidity Willis called lethargy, which the animal spirits could overcome in time by cutting new passages through the brain. As for those who were born stupid, heredity might play a role in Plowmen and Rusticks. "In some Families, reckoning many descents backward," he wrote, "there is scarce one witty or wise man found." He neglected the fact that members of such families also all grew up in the same poor, untutored conditions, which might have had something to do with their so-called stupidity. But he couldn't avoid the puzzling fact that even scholars sometimes had idiots for children—perhaps, he speculated, because they studied too hard and distracted their rational soul. Willis, like many other natural philosophers of his day, believed that a father's soul traveled from the brain to the testicles to create semen; any disruption in the supply could harm a child. Although he saw no cure for stupidity, he believed some of its victims could be helped in special schools, where they could sharpen their spirits as much as possible.

For years Willis hoped to put this mass of ideas on paper—to present the world with what he called a doctrine of the soul, a "psychologia." The word had been in circulation for almost a century. Originally, it referred to the study of Aristotle's soul, the thing that distinguished the living from the dead. The proto-psychologists of the 1500s studied flowers and grasshoppers along with the human mind. By the time Willis used the word "psychologia" (or "psycheology," as it was translated in the English version of his book), its meaning had mutated into something new. By psychology Willis meant an account of the workings of the human soul as a compound of a rational, immaterial substance nested within a swarm of chemical spirits traveling along pathways through the brain and into the nerves. He believed his psychology would prove to be "unconventional and unestablished." In a 1668 pamphlet, he promised that it "will perhaps sometime emerge into the light."

Grief blocked his progress, though. His wife, Mary, grew ill with tuberculosis, taking to bed with heaving coughs. Willis was distraught. Although he had watched thousands of people die from all sorts of diseases, he had never grown hard to their suffering. He was intimate with the grief that followed their deaths: his parents had died of fever when he was twenty-two, and now, at age forty-eight, half his own children were dead. He would make any sacrifice he could to save Mary's life. He offered to quit his wealthy London practice. He would give up all the prestige he had earned and take Mary away from the smoke of London, perhaps back to one of their country houses or to the farm outside Oxford where he grew up. But Mary Willis refused. She had fought the Puritans in her own home, and now she would fight her disease. For years she struggled against the tuberculosis until she died in 1670.

Willis was left in an unbearable abyss. The nights in his house in St. Martin's Lane were long and silent. He began to write down his doctrine of the soul and spent the next two years working on it. "After the death of my dear wife," he wrote at the start of his book, "being lonely, with frequent and unseasonable Studies, that I might the less think on my grief, I have at last finished this, according to my slender capacity." He entitled it *Two Discourses Concerning the Soul of Brutes.*

It was in many ways Willis's greatest work. Combining his

research in anatomy and chemistry with his observations as a doctor, Willis created a material explanation of the soul and its disorders. He did not try to reduce the psychological life to simple mechanics but tried instead to find a pattern of chemical events complex enough to match the complexity of people's inner lives. It was the most complete account of a brain-based psychiatry since the Greeks began practicing medicine, and it foreshadowed modern psychiatry in many ways.

Willis knew that *The Soul of Brutes* was even more dangerous than his previous books. He had transformed the traditional three-part soul, which had existed since Plato, into the corpuscular chemistry of the nervous system. The soul was not just moved to the brain but limited to it, and only through the nerves could it experience the world. Willis had abandoned Descartes's careful compromise, handing even more of the rational soul's work to a material soul. Moreover, this powerful sensitive soul was now no longer a unitary, undifferentiated thing but a collection of spirits that traveled through networks of pathways, a soul that could become diseased like any other part of the body. Once again, Willis wrapped his book in promises that he was not an atheist, that what might seem like a subversive book actually championed the church. Once again he sought protection by dedicating the book to Archbishop Sheldon. He called on his powerful patron, "for my greater safety, to extend your help to me."

Willis's anatomy of the soul was in many ways a picture of his treasured old England transformed into a human body. The king and his royal servants became the rational and sensitive souls, able to ensure the health of the body. But they could do so only as long as spirits flowed peacefully along the nerves and stayed in their proper places in the hierarchy. Just as God appointed a king to rule the nation, He had provided each human body with a rational soul to rule it. With a well-ordered body politic, one could rationally go to heaven. For Willis's generation, the English Civil War had been a fit of national madness in which the king had been killed and the country torn apart. In his secret church, Willis had listened to sermons on the God-given hierarchy of both the country and the church, and at the Restoration he was thankful that England had finally emerged from its insanity with a new king at its head. Yet this mental royalism was nothing but nostalgia.

Willis could see for himself that England's new king was far from a purely rational creature. He was a mercurial, sometimes cruel sensualist. Nor could Charles II rule England the way Willis thought the rational soul ruled the body. After the Civil War, English kings would be able to hold onto power only through crafty diplomacy and compromises with Parliament and the church. England would never be the same again, nor would humanity's understanding of its own soul. Willis had laid down the foundations of a new science which would evolve in the centuries that followed into something that would have both thrilled and frightened him. His "psycheology" was not a scheme of what England was becoming but a dream of what it had been, a dream from his childhood, before the fever of civil war destroyed his family and his country, when he walked across the meadows with the towers of Oxford in the distance.

—

To a twenty-first-century reader, it can be difficult to understand how *The Soul of Brutes* didn't get Willis into serious trouble. He brought the new science into the recesses of the soul, carrying the ideas he first presented in *The Anatomy of the Brain and Nerves* and *Cerebral Pathology* to an extreme. Demons and religious visions had no place of privilege in Willis's psychology, except perhaps as masks of diseases. Although Willis granted man a rational soul that was immaterial and eternal, he buried it deep within the brain, a prisoner of its fleshy structures and weaknesses, relying on an unreliable sensitive soul for its connection to the outside world, vulnerable to madness and idiocy. He made the brain preeminent in the body, but he scattered its ruling tasks—to keep the heart beating, to remember the world, to learn—across a neurological landscape. The brains of animals had a remarkably similar landscape, which meant that their souls were, up to a point, just like our own. They were capable of learning and remembering, of practicing the science of brutes.

Moreover, some of the church's old enemies embraced Willis's ideas. As soon as *The Soul of Brutes* was published, it was pulled into a bitter debate over witches. The clergyman Joseph Glanvill had been spending years carefully documenting witchcraft and ghosts for the Royal Society. He investigated the ghost of a drummer that raised a ruckus in the cellar of a nobleman's house. He described how a murdered corpse suddenly bled when its murderer was forced to touch it, how victims of witchcraft vomited up pins and nails. Here was evidence, Glanvill claimed, of immaterial spirits navigating their way through the universe of mechanical matter. It was not the spiritless world of Hobbes. It was not the self-sufficient, soul-infused world of van Helmont or Paracelsus. It was the mechanical yet spiritual cosmos that many in the Royal Society believed was the proper world for the church.

In the mid-1670s Glanvill was attacked by a critic who hadn't been heard from for twenty years—John Webster, the Puritan surgeon and minister who had called for the destruction of Oxford in 1653. Back then, when the kingdom of God seemed to be at hand, Webster had dreamed that England would be rebuilt from its foundation. Like other radicals, he was quickly disappointed by Cromwell. Webster was so disillusioned that he abandoned organized religion altogether and retreated to Lancashire, where he set up a medical practice. After the return of Charles II, Webster stayed busy with alchemy in his private laboratory, publishing a guide to metalworking. He belonged to the Restoration's great army of silenced dissenters, but Glanvill managed to rouse him to write a tract called *The Displaying of Supposed Witchcraft.*

Glanvill, Webster declared, had been fooled by the workings of the brain, and he invoked Willis as evidence. Webster argued that the widespread belief in witches was due to the very nature of the human brain, in which an evil education could create "a most deep impression of the verity of the most gross and impossible things." He didn't doubt that people sometimes saw specters, but what they saw was not something outside nature. Souls were not separate from nature, he explained, but were part of everything in creation. Soulful nature had fooled Glanvill into believing in witches.

Willis, Webster pointed out, had demonstrated that all people were born with two souls, one immaterial and the other made of particles of spirit. Willis had even speculated that sometimes violent passions could drive the fine particles of the spirit out of the brain and skull and create a kind of "aetherial man" that could be mistaken for a ghost. Webster suggested that when people died, their immaterial, rational soul left their body immediately, while the sensitive soul—a material vapor that suffused the body—would slowly blow away like a cloud in the breeze.

Webster had to struggle to find someone who would give him a license to print his book. He complained to a friend that clergymen blocked a license from the church censors because "I have attributed too much to natural causes." Webster turned to the Royal Society, which had its own imprimatur, but only with the help of his virtuosi friends did it finally publish his book. It would be among the last Paracelsist books published in England. The mechanical philosophy had finally eclipsed all its rivals.

Yet Webster did not taint Willis. In spite of his debt to Paracelsus and Paracelsists such as van Helmont, Willis held onto his Restoration respectability. His brain, his nerves, and the sensitive soul they housed all fitted comfortably enough into the mechanical philosophy. No one worried that Willis was trying to fill nature with mystical, autonomous souls or that he was trying to read their signs by magic. Nor did Willis lurch too far in the other direction and make nature a self-sufficient machine. That is how he avoided the fate of Thomas Hobbes, whose theory of the mind was also deeply rooted in mechanical matter.

In 1660, when Charles had traveled through London to his coronation, Hobbes had been standing among the crowds lining his route. The old man tipped his hat to his student, and Charles stopped to greet him. The bishops around him were appalled; Hobbes was an enemy of the church and the most dangerous atheist in England. Charles ignored their clucks and welcomed Hobbes to court, even giving his old tutor a comfortable pension of £100 a year. Charles enjoyed watching him pit his wit against the clergy. When Hobbes entered the court, the king would announce, "Here comes the bear to be baited!"

The bishops did not disappoint Charles. They attacked Hobbes mercilessly, making him responsible for the libertine atheism that frightened them so much—and for which they dared not blame their own king. "Most of the bad Principles of this Age are of no earlier a date than one very ill Book, are indeed but the spawn of the Leviathan," one critic wrote.

For many churchmen, Hobbes confirmed all their worst fears about the new science. From his pulpit, Stillingfleet declared that to accept the mechanical philosophy was "to suppose Man to be a mere Engine, that is necessarily moved by such a train and series of causes that there is no action how bad soever that is done by him, which it was any more possible for him not to have done, than for the fire not to burn when it pleases."

To accept all this would mean the end of Christianity itself. "If this be true," Stillingfleet warned, "farewell all the differences of good and evil in men's actions; farewell all expectation of future rewards and punishments: Religion becomes but a mere name, and righteousness but an art to live by."

The Oxford circle also tangled with Hobbes during the Restoration. When Boyle published his experiments with his air pump, Hobbes attacked him. Experiments proved nothing, he argued; all kinds of hypotheses could account for the things Boyle saw in his glass receiver, and Boyle could not know which was true. Hobbes argued that instead of experiments, a philosopher should rely on pure deduction from first principles. Hobbes's own reasoning had led him to conclude that there could be no such thing as a vacuum, and he refused to believe any evidence to the contrary.

When Christopher Wren drew up a list of potential members

of the Royal Society, Thomas Hobbes was not among them. The virtuosi claimed they believed in free and open debate, but Hobbes did not abide by their ground rules either for experiments or for the nature of God. It was bad enough that Charles II found Hobbes amusing and listened to him insulting the Royal Society. The virtuosi could not afford to let this troublemaker in their midst. Robert Boyle, so reluctant to speak harshly of anyone, went out of his way to attack Hobbes, even when he was reporting how he cut the heart out of a frog. The frog jumped and swam for another hour, and when Boyle pressed its chest and belly, he could make it croak. "How this experiment will be reconcil'd to the doctrine ascrib'd to Mr. Hobs," he wrote, "or to that of the Aristotelians, who tell us, that their master taught, the heart to be the seat of sense (whence also, though erroneously, he made it the origin of the nerves) let those that are pleased to concern themselves to maintain all his opinions, consider." By exposing Hobbes's ignorance about the brain, Boyle hoped to prevent him from creating any more atheists.

But the virtuosi learned that the bear could maul them if they got too close. John Wallis once claimed that Hobbes wrote *Leviathan* as a way to please Cromwell, conveniently overlooking the fact that he himself had cracked codes for Parliament. Hobbes brought up this embarrassing cryptographical chapter, describing how Wallis "thereby delivered his Majesty's secrets to the enemy and his best friends to the scaffold and boasted of it in your book of arithmetic, written in Latin, to all the world, as a monument of your wit." Wallis never brought up Hobbes's politics again.

As the years passed, Hobbes became more symbol than man. In 1666, bishops blamed the fire and plague on his atheism. Parliament investigated him for blasphemy for two years, and Hobbes desperately burned all the papers he thought might be used against him. He escaped being charged as a heretic, but he was forbidden to write ever again about human nature. Hobbes's books continued to sell well, and a growing number of educated Europeans accepted at least some of his ideas—that people competed by their nature for the things that made them happy, that the state was based on an agreement by its citizens to be ruled. Hobbes managed to enjoy his old age—at seventy-five he was still playing tennis to soothe his tremors and even practiced the lute—but he was intellectually leashed. Trying one last time to make the world understand him, Hobbes wrote an account of the English Civil War, which he argued had been caused by misguided Puritans. But in 1679, Hobbes's final year of life, Charles decided the book was too dangerous and refused to let him publish.

Both Hobbes and Willis accounted for a vast amount of human nature in the language of matter. Unlike Hobbes, though, Willis still left a place in his psychology for an immaterial, immortal rational soul. Mechanical philosophy could explain only so much. It failed to account for the human ability to conceive of oneself or of God or of anything not immediately before the senses—strengthening religion wherever it fell short. We may look back at Willis's neurology and see the subversive seeds planted within it, but at the time they remained unsprouted, allowing Willis to seem like a true conservative, rather than a messenger from Satan or a nature-worshiping pagan.

Indeed, Willis was embraced by both the church and the Royal Society. The *Philosophical Transactions* of the Royal Society mentioned Hobbes's books only to sneer at them; John Webster's book on witchcraft it ignored altogether. But it trumpeted the publication of *The Soul of Brutes,* praising "the learned author of this difficult argument." Charles II might not have liked Willis much, but his royal physician, Walter Charleton, relied on *The Soul of Brutes* when he carried out his own dissection of the passions. In *The Gentleman's Companion,* a popular advice book, William Ramesey urged his reader to consult Willis's *The Anatomy of the Brain and Nerves* "if thou wilt study thy own frame of body." To live properly, a gentleman had to become a neurologist—he had to understand the workings and the weaknesses of his own brain.

Ramesey warned his readers to avoid the ways of Quakers and other enthusiasts. They were not so much sinners as sick.

"They are mad, and therefore have more need of a cure, than such as are in Bedlam," he wrote. Willis provided the perfect anatomy for this sort of condemnation, explaining that "superstition and the despair of eternal salvation" had the same effect on the brain as jealousy and unrequited love, but unlike a physical lover, salvation was the sort of abstraction that only the rational soul could understand. Its yearning for the spiritual object of affection was impressed on the corporeal soul, stirring it into a hot passion and distracting it from its proper work. Driven from their normal paths, the spirits might plunge into fits of raving. In this delirious state, the sensitive soul might conjure up images of angels and demons. And since the rational soul depended on the sensitive soul for its perceptions, a person could be "wholly perverted from the use of right reason."

Priests and bishops happily folded Willis's teachings into their theology. They made casual references to the corpus callosum in sermons on the Resurrection. They no longer saw speaking in tongues as a matter of possession by God or the Devil. It dwindled away to a physical disorder of the brain that gave its victims disordered perceptions of the real world. They would leave religious madness to brain doctors like Willis.

Chapter Eleven

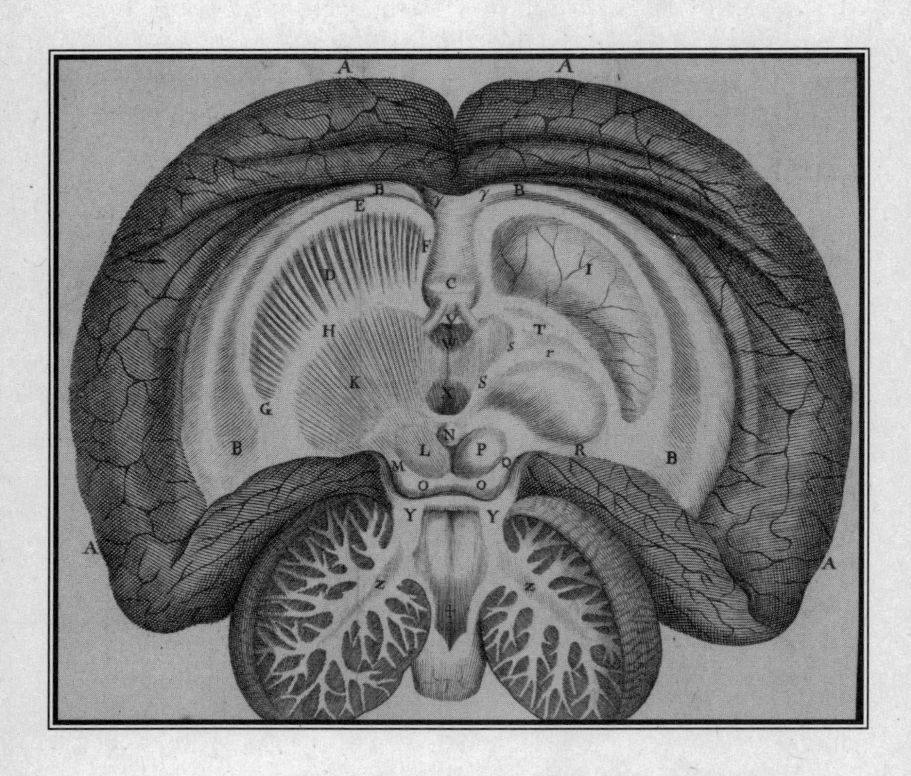

An illustration of a dissected brain from *Two Discourses Concerning the Soul of Brutes*, displaying the unprecedented detail of Willis's anatomy of the brain's innermost regions.

The Neurologist Vanishes

*A*s a young man in 1643, Willis had shouldered a musket to defend a foul, sickly town under siege. In 1673 he would sometimes take up a gun again, but now he trained it on snipe instead of Parliament's soldiers. When he managed to bring down a bird, his servant carried it home for him over the moat that encircled the mansion he now leased twenty miles upriver from London. After his day of hunting, Willis dressed for dinner, which he shared with his new wife, Elizabeth Calley, a woman of good royalist stock. In the evenings Willis would write, but now about drugs instead of psychology.

Willis's cures had become legendary. Some of them he still kept secret, even as countless physicians and apothecaries tried to extract the recipes from him. (One physician promised, to no avail, that he would be "as secret as though I were its custodian in the shrine of Hippocrates.")

Willis set out to organize the remedies he had developed over thirty years into a single book. Instead of a simple catalog, he created a rational science of drugs, based on corpuscles and anatomy. He saw it as a final triumph over the Paracelsus-inspired radicals who had perished in the plague a few years earlier. Willis believed that only if physicians learned the "mechanical means of the working of Medicines in our Bodies" would it be possible to fight off the "ignorant Pseudochymists" and the "vilest scum of the people who bark against and fling dirt upon Physick." He set out to write the book that the Civil War had stopped William Harvey from writing. Willis was sure that "physick being perfected in all its parts, may grow to a true science and be practiced with greater certainty not inferior to the Mathematicks."

Rational Therapeutics, as Willis called it, was the last book he would write, and it was a fitting cap to a paradoxical career. Willis believed to his death that Galen's remedies were real cures, and he used the new language of corpuscles to give them a new authority. Bloodletting worked, he claimed, because blood was a vital flame that sometimes had to have sulfurous particles vented from it, "as oil from a lamp." Galen's prescription for vomiting was effective because it removed the heavy phlegm that built up in the folds of the stomach, leaving the stomach "cleansed as with a broom." Younger physicians such as Richard Lower were now backing up their claims with experiments, but Willis did not bother. He simply presented his wonderfully imagined stories as fact.

Despite its shortcomings, *Rational Therapeutics* also displayed Willis at his best: as a sharp-eyed anatomist and an extraordinarily observant physician. With the help of Edmund King, his anatomical assistant for *The Soul of Brutes,* Willis made an unprecedented study of the stomach, lungs, and other organs. He realized that sweet-smelling urine was a sign of diabetes. He distinguished between acute and chronic tuberculosis. He described emphysema for the first time. He analyzed diseases and medicines chemically. He scorned horoscopes in medicine as "altogether frivolous."

Willis became so ill while he was writing *Rational Therapeu-*

tics that a rumor spread that he had died, but he recovered and went back to his business. He added a three-thousand-acre estate to his holdings and started the second volume of his book. A few months later he developed a nagging cough, which turned to pneumonia in the fall of 1675. Edmund King served as his physician, but both Willis and King knew after a few weeks that no cure in *Rational Therapeutics* would save him.

Willis approached death as he had always hoped he would, his rational soul still in charge. He made some minor changes to his will, in which he left hundreds of pounds to charity, summoned his brother-in-law John Fell to say good-bye, took Holy Communion, and commended his soul to God.

Fell arranged a funeral that Willis's grandson later described as "ridiculously sumptuous." He was buried in Westminster Abbey, and in the months that followed, ministers and philosophers sang his praises. The Royal Society said that his death "may certainly be reputed a public loss, and a detriment to the very Faculty of Physick."

The poet Nathaniel Williams wrote an elegy that summed up the age's opinion of Willis.

Thou knewst the wondrous art,
And order of each part,
In the whole lump, how every sense,
Contributes to the health's defense.
The severall Channels which convey,
The vitall current every way,
Tracks wise Nature everywhere,
In every region, every sphere,
Fathomest the mistery
Of deepe Anatomy.
The unactive carcasse thou hadst preyd upon,
And stript it to a sceleton,
But now alas! The art is gone,
And now on thee
The crawling Worms experience their Anatomy.

—

More than any other individual, Thomas Willis ushered in the Neurocentric Age. He did for the brain and nerves what William Harvey had done for the heart and blood: made them a subject of modern scientific study. His mixture of anatomy, experiment, and medical observation has set the agenda of neuroscience into the twenty-first century.

In redefining the brain, Willis redefined the soul as well. It was banished from the liver and the heart, restricted now to the brain and nerves, where invisible corpuscles produced emotions, memory, and perceptions. Willis still saved a place in the brain for an immaterial, immortal rational soul, but one that depended profoundly on the sensitive soul. It could not perceive the outside world on its own, and the diseases of the sensitive soul could eclipse the rational soul as well. And to cure the soul, Willis tried to alter the corrupted corpuscles of the brain, perhaps by dancing and jesting or perhaps by a drink of steel syrup.

Willis's doctrines of the brain and the soul became part of the bedrock of modern Western thought, and they still lurk beneath many of our beliefs about ourselves today. And yet, despite his lasting impact, Willis's reputation began to fade soon after his death. By the eighteenth century, he was no longer remembered as a great doctor with an unparalleled power to see into the soul. Those who remembered him at all were more likely to call him a dreamer seduced by his own speculations. Willis still lies buried in Westminster Abbey, but visitors today rarely bother to look down as they walk over his stone-carved name. Resting in a crowd of kings and poets, Willis lies in as much obscurity as Anne Conway in her unmarked village grave.

One man bears the lion's share of responsibility for why few people today know about Thomas Willis: a certain eager student of his named John Locke.

When Locke came to Oxford in 1652, he showed none of the glaring genius of his fellow students Christopher Wren and

Richard Hooke. The virtuosi took no notice of him for years, never mentioning his name in their journals and letters. The university lectures on Aristotle bored Locke, who amused himself by reading romances and writing flirtatious letters to women he never actually pursued. He developed an amateur's curiosity about medicine, filling notebooks with recipes that called for hedgehog grease and carved-up puppies. He once took the heart out of a frog and watched the animal leap about until it died. Locke did these things more to pass the time than with any pretense of practicing the new science.

By the end of the 1650s, however, he began to transform himself from a dilettante to a natural philosopher. He read Descartes, Gassendi, and other great philosophers. He began to read hundreds of medical books, diving into Willis's tract on ferments as soon as it was published. He became friends with Richard Lower, who introduced him to the experiments he and Willis were carrying out on blood. Along with Lower and Anthony Wood, Locke took chemistry lessons from one of Boyle's assistants. Wood complained that "this John Lock was a man of turbulent spirit, clamorous and never contented. The club wrote and took notes from the mouth of their master who sat at the upper end of a table, but the said J. Lock scorn'd to do it; so that while every man besides of the club were writing, he would be prating and troublesome."

Locke was incapable of playing the passive student, but he was apparently willing to sit through at least one official lecture at Oxford—that of Thomas Willis. Locke attended Willis's lectures twice a week, listening to the doctor paint a picture of the soul lodged deep within the flesh of the brain, of the spirits traveling along the nerves from the eyes and other senses to the brain, where they projected their images for the rational soul to see and formed impressions on the cortex. Locke was so impressed by Willis's lectures that he went to Richard Lower, borrowed his massive stack of notes of his master's lectures, and copied them, filling up hundreds of pages in his notebooks in the process.

Locke earned a place in the Oxford circle, attending its weekly meetings and carrying out experiments on its agenda. He became an expert on plants, traveling into the countryside to collect thousands of specimens which he pressed in the Latin exercise books of the students he tutored. Locke analyzed them in a little laboratory he set up in his rooms with the physician David Thomas, where they burned the plants down to oils and phlegms. Locke tried to make the alkahest, the legendary substance that van Helmont had claimed could turn anything to the water from which it originally came. He investigated blood, distilling it, extracting salts from it, theorizing that it turned red when it absorbed some substance from the air—an idea confirmed a few years later by Lower.

When Locke wasn't carrying out medical research, he was speculating on the hidden workings of the body. He built an eclectic philosophy as he browsed different schools of thought for ideas he found reasonable. He liked Descartes's particles, jostling their way through pores and sieves, but he doubted that they could create the full complexity of life on their own. How, Locke wondered, could Descartes's simple sieves allow both mint and marjoram to grow from the same water? Only some kind of living seed, some kind of ferment, could be responsible—not a soul or spiritual messenger, but "some small subtile parcels of matter which are apt to transmute far greater portions of matter into a new nature and new qualities."

Locke might have carried on these speculations if he had stayed in Oxford, but in 1666 his life suddenly changed when he met an ailing aristocrat named Lord Ashley. Ashley had come to Oxford hoping to ease a chronic pain in his side by drinking Richard Lower's healing waters at Astrop. He arranged with David Thomas to have a dozen bottles brought to him in the comforts of Oxford. But just before Ashley arrived, Thomas had to leave town and begged Locke to take care of the errand for him.

Locke apparently forgot about the water and had to make some awkward apologies to Ashley when they met. Despite the bad start, the two men immediately struck up a friendship, and the following day they were sitting side by side, chatting away, as they both sipped

Astrop waters. Ashley found Locke an intriguing thinker and enjoyed talking to him about politics and religion. Their friendship grew over the next few months, until Ashley invited Locke to come to his home in London and serve as his physician.

Ashley was one of the most ambitious politicians in all London. He had supported Charles I in the Civil War until 1644, when he switched over to Parliament's side. He then became powerful in Cromwell's Commonwealth, but as a Presbyterian, he fell out of favor with the leaders of the army. By the end of the 1650s, he was secretly conspiring to bring Prince Charles back to England. Now his star was rising again in Charles's court, although given his past, no one was sure where his loyalties lay.

Of the many people Locke met in Exeter House, Ashley's sprawling mansion, the most influential was a gruff, sad-faced physician named Thomas Sydenham. Sydenham was a schoolmate of Willis's at Oxford, but they had little in common. Willis came from royalist stock, while Sydenham grew up in Puritan Dorset. Willis fought for Charles I, while Sydenham fought for Parliament twice—against Charles I and then against his son. Like Willis, Sydenham chose medicine, and he too became disillusioned with Aristotle. (A friend recorded how Sydenham declared that "one had as good send a man to Oxford to learn shoemaking as practicing physick.") But Sydenham steered clear of Willis's circle. He found their fascination with invisible corpuscles and hidden anatomy too speculative and struck out on his own. He preferred patterning his medicine after Hippocrates, relying on careful bedside observation of patients.

Sydenham left Oxford for London in 1655 to join his two brothers, who had risen to the highest ranks of Cromwell's army, but when Monck took control in 1660, his network of powerful Commonwealth friends vanished. He began to spend most of his time tending to patients from the lower classes, "those poor people whom my lot engages me to attend," as he later wrote.

As Sydenham saw patients by the hundreds, he began to see patterns. The same clusters of symptoms took a nearly identical course through different people. Galenist doctors treated a disease as the unique disturbance of an individual, but Sydenham now saw it otherwise. "The selfsame phenomena that would be observed in the sickness of a Socrates you would observe in the sickness of a simpleton," he declared. A wave of "intermittent fevers," probably a virulent strain of malaria, swept through his neighborhood in the early 1660s, and Sydenham found that no matter whom it claimed, it followed the same distinct and specific course. He found ways to distinguish other fevers, building up lists of linked symptoms that separated one disease from another. "It is necessary that all diseases be reduced to definite and certain species, and that with the same care which we see exhibited by botanists," he wrote. He classified diseases as if they were thistles and lilies.

Sydenham's huge practice also allowed him to gauge how well different kinds of remedies worked. He recognized that the poor sometimes overcame their illnesses without any medicine at all and that medicine even could do more harm than good. Bad medicine, Sydenham claimed, had caused more slaughter and havoc "than hath bin made in any age by the sword of the fiercest and most bloody tyrant that the world ever produced." Sydenham didn't know how nature healed people, but he was willing to accept his own ignorance rather than invent some ornate explanation. "Nature by herself determines diseases, and is of herself sufficient in all things against all of them," he declared. The best a physician could do would be "joining hands with nature."

He experimented with different treatments and discovered that his patients with smallpox fared better when he let them wear thin clothes and rise from bed; the traditional practice would have required keeping them warm and giving them cordials to drink. It did not bother him that he did not know exactly why his remedy seemed to work. It would make no more sense to expect a master cook to explain the chemistry of his stews.

Other physicians were outraged and tried to have his license revoked, but Sydenham would not change his medicine. "It is my nature to think where others read, to ask less whether the world agrees with me than whether I agree with the truth; and to hold cheap the rumor and applause of the multitude."

Locke was fascinated by Sydenham's radical medicine and began to follow him on his rounds. As he treated Lord Ashley's household, Locke began to imitate Sydenham. One night in 1668, Ashley began to vomit violently, and his skin turned rusty red. After a few weeks of paroxysms, a tumor pushed up under the skin around his liver. It was the size of an ostrich egg and yielding to the touch. Locke had Ashley's skin seared open and then had the tumor ruptured. In addition to the normal pus and blood, a baffling collection of what Locke called "bags and skins" emerged from the incision. To let the wound drain, Locke had a silver pipe placed in Ashley's side.

To understand the bags and skins, Locke followed Sydenham's example. He polled all the physicians he knew for any similar cases they might have seen and collected tales of people who had had the same condition and recovered. (The bags and skins were actually the cysts of a tapeworm.) After gathering what evidence he could, Locke decided to leave the pipe in Ashley's side, so that he would not have to be cut open again in later years. It was a decision for which Ashley would be forever grateful to Locke, despite the nickname it supplied his political enemies: Tapski.

For Locke, Sydenham was the breathing opposite of his Oxford professors: a brash, practical healer interested only in saving lives through direct experience. Locke peppered his new friend with fundamental questions about his medical methods. Could Sydenham actually know *anything* about the inner causes of diseases or the workings of a healthy body? Did he even need to know in order to be a good physician? Before long, the two men's thoughts merged, like questions and answers batting around within a single skull. Sydenham would sometimes start writing an essay on medicine and Locke would finish it; Locke sometimes took dictation, editing Sydenham along the way.

Locke and Sydenham were exploring a sort of medicine very

different from that of Thomas Willis, whose new London practice was attracting the rich and powerful. The diarist John Ward recorded in his journal that "Sydenham and some others in London say of Dr. Willis that he is an ingenious man but not a good physitian, and that he does not understand the way of practice."

Willis championed autopsies and microscopes as ways to understand how medicine worked, but Locke and Sydenham believed that these ambitions were pure folly. "All that Anatomie can do is only show us the gross and sensible parts of the body, or the vapid and dead juices," they declared. "The true use of parts and their manner of operation anatomy has hitherto made very slender discoveries. Nor does it give very much hopes of any greater improvement." Anatomy, Locke and Sydenham declared, "will be no more able to direct a physician how to cure a disease than how to make a man."

Willis's attempts to discover the mind by mapping the brain were even more foolish. "The brain is the source of sense and motion," they wrote. "It is the storehouse of thought and memory as well. Yet no diligent contemplation of its structure will tell us how so coarse a substance (a mere pulp, and that not over-nicely wrought), should subserve so noble an end. No one, either, can determine, from the nature and structure of its parts, whether this or that faculty would be exerted."

In Sydenham's company, Locke became convinced that all causes in medicine were beyond human understanding, that physicians could only observe symptoms and grope for cures that worked for reasons unknown and probably unknowable. When Locke discussed philosophy with his friends, a radical question occurred to him. What exactly could a person know about *anything*—not just medicine, but any branch of human knowledge? Could it be that a lot of debate and persecution was a needless confusion brought on because people did not understand the nature of thought itself?

One particularly rich source of confusion could be found in what Locke called "the imperfection of words." He once found himself in an argument over whether nervous liquor flowed in the nerves, an idea championed by Willis but doubted by others. After a while, Locke stopped the dispute and asked his friends what they actually meant by the word. "They at first were a little surprised by the proposal," he later wrote. Thinking about the meaning of words this way was unusual in Locke's day, but his friends quickly realized that they all agreed that some kind of "subtle" fluid passed through the nerves. Locke showed them that all their arguing over whether it was actually a liquor, whatever that meant, was beside the point.

At some point in 1670, Locke decided that he and his friends would have to get past these simple obstacles of language and thought before they could truly understand anything. "It was necessary to examine our own abilities," he later wrote, "and see what objects our understandings were or were not fitted to deal with." Locke took on the task himself, starting an essay that he worked on from time to time, periodically producing what he referred to as "incoherent parcels."

While Locke was working on his essay, Thomas Willis paid a visit to Exeter House. Lord Ashley had used Willis's services in the past and now the doctor came to check up on his health, inspecting his silver pipe and gently probing the hole from which it sprouted. It's possible that Willis crossed paths with Locke during his house call. Willis might have greeted Locke with the friendly condescension of an old teacher to his former student. Perhaps, as Willis probed Tapski's silver pipe, he asked Locke if he had had a chance yet to read his new book on the souls of brutes, which had emerged out of the lectures Locke had heard at Oxford eight years earlier. Perhaps Locke replied with a polite smile that he would be sure to get around to it as soon as he finished working on a little essay of his own.

—

"We should do well to commiserate our mutual ignorance," Locke wrote. He believed that his philosophy had a lot to say not only about the nature of understanding but about how people should live together. A state, Locke declared, "had nothing to do with the good of men's souls or their concernments in another life." It should not try to do anything beyond protecting the natural rights of its subjects. If a person's views were dangerous, the state might be justified in stopping him from publishing those opinions, but it shouldn't force him to recant. No one could be so confident of his own knowledge to have such power over the minds of other people.

Ashley shared many of the same opinions and made Locke one of his chief political advisors. His political star was rising at the time, and by 1672, when he was created the first earl of Shaftesbury, he was one of the most powerful members of Charles II's court. He used his growing power to fight for religious tolerance toward dissident Protestants, and eventually found in Charles II an ally in his struggle.

Shaftesbury had his doubts about the king, though. He suspected that Charles II would use the campaign for religious toleration as a way to make England a Catholic nation. While he and Locke might be tolerant of many religions, they saw Catholicism as a threat to their country's autonomy. For years rumors had circulated about Charles's upbringing in exile with his Catholic mother in France. Shaftesbury was even more suspicious of Charles's brother James, whose conversion to Catholicism was an open secret. Since Charles had yet to produce a male heir, James was next in succession to the throne.

His suspicions eventually turned to action: Shaftesbury began trying to push Catholics out of the government and to prevent James from taking the throne. Suddenly he and Charles had become implacable enemies, and in 1673 Shaftesbury was forced out of the government. But even out of power he kept conspiring. He began to meet regularly with a group of like-minded politicians at the Swan tavern, where they hatched their strategies against the king. They created something new in English history, a network of statesmen and businessmen united not by a religious war so much as by politics—in other words, a political party. Its enemies—the loyal bishops and landowners who sided with Charles—branded it the Whig party (named after the word "Whiggamore," meaning a Scots Presbyterian bandit).

The Whigs were buoyed up by a burst of anti-Catholic terror that seized England in 1678. A renegade chaplain claimed that he had evidence of a plot by Jesuits to kill Charles II and make England Catholic again. Two dozen people were rounded up, convicted of treason, and hanged. The Popish Plot brought Shaftesbury back into power, leading Parliament as Lord President of the Council. He maneuvered in Parliament to block James from the throne, trying to pass a law forbidding the succession of a Catholic king and to get Charles's bastard son the duke of Monmouth declared heir. But Charles would have no bastards lead the house of Stuart. Instead, Charles countered Shaftesbury with a savvy that his father had utterly lacked. He placated Parliament with promises while searching for pretexts to have the leading Whigs arrested.

Shaftesbury was put in the Tower of London and charged with treason, but a grand jury of Whiggish London merchants refused to indict him. Free once more, Shaftesbury decided his only recourse now was war. But Shaftesbury was no Cromwell, and once the Popish Plot was exposed as a fraud, the country lost its hostility to the king. With his rebellion a failure and Charles closing in on him once more, Shaftesbury disguised himself as a Presbyterian minister and fled to Holland in 1682, where he died two months later.

In the months after Shaftesbury's death, England became a dangerous place for Locke. Officially, he was just a minor player in the Whig party, but everyone knew he was Shaftesbury's closest counsel. Locke worried that his writings, all still unpublished, might get him hanged. He had just finished writing down some thoughts about how the people were entitled to overthrow a leader who violated their natural rights, virtually granting permission to a Shaftesbury to wage war on Charles. As Charles began having Whigs arrested in August 1684, Locke burned some of his papers, packed a few boxes, and fled to Holland.

Even from across the North Sea, Locke felt Charles's wrath. He learned that the king had added his name to a list of revolutionaries that England wanted returned from Holland. Locke wrote to powerful friends in England, whining that he had only lived in Shaftesbury's house as a medical advisor, that he had made hardly any money in that role, and that his published writings consisted of a couple of short poems in the 1650s. The last point was true, but Locke was dishonest about the rest. In any case, Charles didn't waver, and so Locke went into hiding under the name Dr. van der Linden. It was as a fugitive that Locke finally had the leisure to finish the book he had been writing in incoherent parcels for over a decade. It would ultimately be known as *An Essay Concerning Human Understanding.*

—

Locke claimed great humility in the undertaking. "It is ambition enough to be employed as an under-labourer in clearing the ground a little, and removing some of the rubbish that lies in the way to more knowledge," he wrote. But Locke was actually doing something unprecedented: he was dissecting the way in which humans acquired knowledge. "The understanding, like the eye, whilst it makes us see and perceive all other things, takes no notice of itself," Locke explained, "and it requires art and pains to set it at a distance, and make it its own object."

Locke's essay was also unprecedented in its style, which was based on everyday experience that anyone could agree was true. "What perception is," he wrote, "everyone will know better by reflecting what he does himself, when he sees, hears, feels, &c., or thinks than by any discourse of mine."

Locke would not bog himself down in the kinds of questions Thomas Willis had tried to answer. "I shall not at present meddle with the physical consideration of the mind," he wrote, "or trouble myself to examine wherein its essence consists, or by what motions of our spirits, or alterations of our bodies, we come to have any sensation by our organs, or any ideas in our understandings, and whether those ideas do, in their formation, any or all of them, depend on matter or no." Locke claimed that the workings of the mind were incomprehensible. He would talk only about ideas themselves and based only on what everyday experience confirmed. He would use what he called a "plain historical method" of gathering evidence, the same method that his friend Sydenham used with diseases.

Locke argued that the mind was empty at birth. "Let us then suppose the mind to be, as we say, white paper, void of all characters, without any ideas," he proposed. "How comes it to be furnished?" He argued that through the senses, ideas entered the mind and furnished it like curios in an empty cabinet. The more often we encountered an idea, the stronger an impression it made on our minds. The idea of a triangle or justice or God seemed innate to our minds only because its truth had been impressed on us so many times, even from before we could remember.

The mind could combine simple ideas and compare them to produce more complex ones, which it could then combine into more complex ones still. But as complex as the ideas might get, human understanding could not reach beyond a narrow compass. "The simple ideas we receive from sensation and reflection are the boundaries of our thoughts," he wrote, "beyond which the mind, whatever efforts it would make, is not able to advance one jot; nor can it make any discoveries, when it would pry into the nature and hidden causes of those ideas."

Philosophers since Aristotle believed they could discover the essence of things, but Locke denied them that power. Words no longer referred to these essences but were merely labels people pasted onto collections of related ideas in their own minds. In both nature and human affairs, Locke declared, we have nothing to go on other than the ideas that have entered our imperfect senses.

Locke did not surrender to the limits of human knowledge, however. As a doctor, he had to make some kind of provisional diagnosis and then act on it, even if he didn't understand a disease at its deepest levels. To refuse to believe anything because we cannot know all things with certainty, Locke said, made as much sense as refusing to walk because we can't fly. Humans should gather knowledge within their compass, aware that they cannot know the essences of things, that they can only find likelihoods.

"Hypotheses," Locke wrote, "if they are well made, are at least great helps to the memory, and often direct us to new discoveries." It was important, though, to frame a hypothesis carefully, only after examining what was known already and perhaps carrying out some experiments, and to be ready to cast it off for a new one that turned out to be more powerful.

Locke's *Essay Concerning Human Understanding* has an ordinary feel today, which is exactly what makes it so historic. Locke absorbed the ideas of Willis, Descartes, Gassendi, Hobbes, and the other philosophers of the seventeenth century and turned them into something new: a psychology of rational thought in which God was peripheral. In human affairs on Earth, Locke now offered complete autonomy. Instead of seeking metaphysical certainty in the Bible or some Greek text, humans had to grope for the probable. Willis and the rest of the virtuosi who emerged from the English Civil War pondered how they should go about gathering knowledge through experiments and observations, but only in an ad hoc way. It was Locke who transformed this kind of thinking into a full-blown philosophy, one that would become the heart of scientific method. Locke's essay feels ordinary only because he changed the world around him.

In February 1689, Locke's exile came to an end. He boarded the *Isabella,* a ship belonging to the fleet of Princess Mary, niece of Charles II and wife of Prince William of Orange. With him he took sixteen boxes, thirteen of which contained books; another held "a little casque with an iron furnace in it." The *Isabella* joined the rest of the princess's fleet and headed west over the winter sea for England. Mary was traveling there to become queen, and Locke to become the philosopher of the new age.

The England to which Locke and Mary were sailing had

changed for good. Three years after Charles II had driven Shaftesbury from the country in 1682, he succumbed to a long-standing kidney disease. Charles was purged, plastered, scalded, and drained of quarts of his blood. Dryden later wrote that the doctors were "prescribing such intolerable pain / As none but Caesar could sustain." Yet Charles somehow remained rational on his final day, still recognizing faces and talking to people, having the curtains drawn back to see the morning light, even reminding his servant to wind his eight-day clock. A bishop who attended the king insisted in the final hours that Charles receive the sacraments. The king grumbled, "He would think of it."

His brother James II took the throne and quickly proved to be an autocrat with little patience for his opponents. After a failed uprising by the duke of Monmouth, he put Catholic officers in charge of the army to guarantee its loyalty. He ignored Parliament as he pushed to suspend the laws against Catholics and dissenters. His opponents began to look abroad for hope. James's daughter Mary was married to Prince William, a staunch Protestant who had led a war against Catholic armies in France and Spain. They pictured William as their king, defending England from the Catholic threat, and Mary preserving the Stuart line of succession. When James had seven bishops arrested for defying his religious reforms, William saw his opportunity to invade. The fighting was brief, and James was soon fleeing London in the dark of night. In a fit of spite, he had the writs for a new free Parliament burned and the Great Seal thrown into the Thames before he left the country. Thomas Willis was lucky not to have lived to see an English king fall so far from the throne of the rational soul.

William and Mary took the throne, agreeing to a long list of terms that could have come straight out of the essays Locke was about to publish. There would be no more standing army, no taxes beyond those raised by Parliament, no special courts prosecuting trumped-up charges. The people would be free to petition, elections would be free and regular, Parliament would meet annually and serve as the main ruling body of England, determining the right of succession. England, which had begun the century ruled by divine right, was ending it with the birth of the modern state.

The *Isabella* docked at Greenwich on February 12, 1689. London had changed since Locke had left for Holland, not just politically but scientifically as well. The explosion of genius that marked the 1660s and 1670s was burning out. The virtuosi had succeeded in establishing an experimental mechanical philosophy, while the rival schools based on Paracelsus's mystical chemistry or Aristotle and Galen were fading away. Anne Conway's all-spirit universe had failed to win a following. Willis had made the brain triumphant over the heart, stomach, and liver. But the dreams of unlimited knowledge and mastery of nature had gone unfulfilled. "'Tis impossible to conceive how so honest, and worthy a design should have found so few Promoters, and so cold a welcome in a Nation whose eyes are so wide open," wrote John Evelyn, a member of the Royal Society. And now the most prominent virtuosi were dying off. William Petty, who never found the political power he was sure his political arithmetic would bring him, was killed by gangrene. John Wilkins never managed to build a chariot that could go to the moon, but he held onto his good spirits even on his deathbed in 1672. His final words were, "I am ready for the great experiment."

Boyle had helped crush Aristotelian accounts of matter, but his experimental alchemy would not mature into a proper science for another hundred years. The microscopes that Hooke and Wren built did not reveal the corpuscles that they believed constituted the universe, exposing instead a series of mysterious landscapes inhabited by equally mysterious creatures. The most spectacular example of the medical potential of the new science, Richard Lower's transfusions, had gone nowhere. Thomas Willis had changed the way people thought about the brain and soul, but when it came to medicine, the outsider Thomas Sydenham was beginning to gain a reputation as the English Hippocrates. Willis meanwhile was beginning to look like someone too easily seduced by his own speculations.

Christopher Wren, now sixty-seven, was beset by regrets and doubts. His dream of a new, rational London had been dead for more than twenty years, and in order to get his scattered buildings erected, he had to grapple with ever-changing politics. In his fits of frustration, he thought nostalgically about his work at Oxford and during the first years of the Royal Society. "He has been often heard to complain," his son later wrote, "that King Charles the 2nd had done him a disservice in taking him from the pursuit of those studies, and obliging him to spend all his time in rubbish; (the expression he had for building) for, had he been permitted to have follow'd the profession of physick, in all probability he might have provided much better for his family."

Wren left behind a trail of half-finished, half-published ideas—great telescopes that were never built, schemes to measure the longitude of the Earth never realized. For years Wren had struggled to finish the great project started by Galileo: to understand how objects moved and collided on Earth and in the sky. He made a few steps on his journey, but it would be completed in 1687 by Isaac Newton.

And so it was that England was ready, both politically and scientifically, for new ideas like the ones Locke had to offer. In quick succession, Locke published his three greatest works, *Two Treatises of Government, A Letter Concerning Toleration,* and *An Essay Concerning Human Understanding.* Together these books would set the tone for the coming century of the Enlightenment. His politics would inspire the writers of the United States Constitution, while his philosophy of knowledge influenced the way scientists investigated nature.

Locke also influenced the way philosophers pondered the mind itself. He dismissed details of neurology and concerned himself with ideas and how they fit together, and generations of philosophers followed his lead. It would take neurologists 150 years to show that Willis was right, that studying the anatomy and chemistry of the brain can indeed reveal the workings of the mind, that they can map the geography of passion, reason, and memory.

But for all Locke's skepticism about his teacher's project, Willis's neurology runs through *An Essay Concerning Human Understanding* like buried steel beams, invisibly bearing the weight of Locke's philosophy. Our understanding is limited by our anatomy—an anatomy which Locke quietly borrowed from Willis's dissections. Ideas, Locke wrote, come to the mind from the outside world through the nerves, which he called "the conduits to convey them from without to their audience in the brain, the mind's presence-room." None of the ordinary experience that Locke so treasured could have revealed this to him. Locke's isolated mind, its blank sheet of paper, and countless other details already existed in the lectures that Willis delivered to him in his student days.

It's true that Locke preferred to speak only of ideas, but it's sometimes hard to tell the difference between them and Willis's spirits. In the healthy brain, Locke wrote, ideas are formed as "trains of motions in the animal spirits, which once set a going, continue in the same steps they have been used to; which, by often treading, are worn into a smooth path, and the motion in it becomes easy, and as it were natural." He used the same images that he had heard in Willis's lectures twenty years earlier, as his teacher explained how madness was caused when spirits cut new paths through the brain, "in which, whilst they flow, they procure unaccustomed notions, and very absurd." Willis believed these altered spirits presented the rational soul with a false image of the world. Locke likewise believed that ideas joined together in wrong ways in the madman's mind.

Locke was giving his old teacher a compliment that was at once backhanded and profound: Willis's brain had become an agreed-upon fact, one that didn't even need pointing out.

—

The Neurocentric Age would evolve into something far beyond what members of the Oxford circle thought they saw in the brains they dissected in Beam Hall. But Robert Boyle, at least, seems to have had a glimpse of the future just before he died in 1691.

Boyle spent the last months of his life in pain, both from a stroke he had suffered twenty years earlier and from the torments of his conscience. Knowing that his end was coming, he spent his final year writing and rewriting an enormous will, which he hoped would redeem him from his sins—particularly the ones he didn't know about but was sure he had committed. Boyle left some of his alchemist's recipes to Locke and Newton to keep safe, urging them never to make them public and fretting that he might have unknowingly consorted with the Devil in his time as an alchemist. He left money to the Irish families whose land had been taken over by his father and for an annual lecture to demonstrate the falseness of atheism.

The threat of atheism worried Boyle more and more as he grew older. In his last few books, such as *The Christian Virtuoso*, he attacked the atheism of Hobbes's unsupervised atoms and championed the immaterial rational soul. The rational soul can imagine things that no arrangement of matter can imagine—not only God but even the simplest sort of mathematics, he wrote. The rational soul he called "a kind of imprisoned angel."

Over the course of his life, Boyle had made himself the "character of a Christian philosopher," in the words of one bishop. Joseph Glanvill liked to say that in another age Boyle would have become a "deified Mortal." Despite the praise, despite all the books Boyle wrote, he could not tame his conscience. Six months before his death, Boyle was so desperate that he asked his old friend Edward Stillingfleet (now a bishop) to see him. When the two men met, Boyle handed the bishop an anxious list of notes and questions he had put together. Boyle had decided early in life that taking vows or oaths was a sin, and he worried that he might have somehow taken one without realizing it. Maybe when he promised to do a favor for someone, he unwittingly took a vow? He begged Stillingfleet to help him sort out the fuzzy, thirty-year-old memories that might be hiding a nest of sins.

Stillingfleet tried to soothe him by explaining that sometimes our brains did things that we couldn't be held accountable for.

"Flashy emanations," as the bishop called them, could jump into our mind and come out of our mouth without our realizing what we were saying. The bishop used the same neurological balm when Boyle asked whether he had committed the sin against the Holy Spirit—whether he, even for an instant, had doubted the reality of God and His work. Stillingfleet reassured him that if he was afraid of committing a sin, he could not have committed it, even if the thought passed through his mind. Thoughts just sometimes did that. Fevers and other physical ills could drive blasphemous ideas through the brain, but God would not write them down in His account book of sin.

Apparently Boyle didn't feel relieved after Stillingfleet's visit, because a week later he was firing the same questions at Bishop Gilbert Burnet, the clucking observer of Charles II's weaknesses. The meeting with Burnet was even more disturbing. Burnet promised that his accidental oaths must be false memories created by his own brain. A man who worried too much could easily "raise or excite in himself a suspicion and father it upon his memory," Burnet said.

Boyle then brought up the blasphemy of doubt. Burnet laughed it off. "In our Age," as he proudly called it, that almost never happened. The only way to commit the sin against the Holy Spirit would be actually to witness a miracle, recognize that it was true, and then deny it. Having inner doubts simply didn't matter anymore. After all, everyone knew that such doubts were not moral failings of the soul but just "mere effects of distempers of the body or the brain and were mechanical effects, and as it were schemes of ideas of thoughts that arose in the mind from the kinds of distempers of the animal spirits without any cause manifest to us."

Boyle repeated Burnet's and Stillingfleet's words later to his secretary, who scribbled them down. As he listened to himself transforming sin into the wayward movement of animal spirits in language that could have been extracted from Willis's books, which gave his agonizing struggle for God's love the same status as a fever, and thought about how he himself had acted as midwife to "our Age," in which the brain had such a mind of its own, perhaps Boyle wondered if, of all his unwitting sins, this was the greatest.

Chapter Twelve

"CREDULITY, SUPERSTITION, AND FANATICISM, " BY WILLIAM HOGARTH (1762). THE "BRAIN-THERMOMETER" IN THE LOWER RIGHT CORNER IS ALMOST CERTAINLY ADAPTED FROM CHRISTOPHER WREN'S ILLUSTRATION IN THOMAS WILLIS'S *THE ANATOMY OF THE BRAIN AND NERVES.*

The Soul's Microscope

To understand just how important those summer days in 1662 were when the Oxford circle gathered at Beam Hall to huddle around a headless corpse and its stripped skull, to cradle a scooped-out brain and follow its courses of blood and spirits to the sources of joy and madness, you have to leave their company. Leave Beam Hall and rise above the tips of Oxford's cathedral towers. Soar west and head out over the North Atlantic. Move forward in time, as three-masted ships set out for the New World, to be replaced by steamships trailing smoke, to be replaced in turn by container ships and cruise liners as the western horizon grows a stubble of America. Make landfall on what was once a colony King Charles gave his brother, James, a wedge of pine scrub and oak hills called New Jersey. A university rises in the village of Princeton, brick and clapboard turning to concrete and glass. Pass through the front door of Green Hall, descend to its windowless basement, pass into a dim room, and come to a stop before a young man lying on a slab. Unlike the corpse you left behind at Beam Hall, this man still breathes. His brain has not left his skull.

The young man lies in a device known as a magnetic resonance imager. A giant white ring encircling his head houses a massive magnet along with tubes of liquid helium that pulse like a soft heartbeat. The young man's body is loosely corseted by straps. Wires hang from two of his fingertips; two other fingers rest on a pair of buttons. Hollow plastic tubes snake up his side and slip into his ears. He is alone in this room, but if he lifts up his head, he can look down past his sneakers and through a window into an observation room, where people are working. Two of the people roll from computer to computer in chairs on casters. The third, a twenty-eight-year-old man with dark coppery hair in tight curls, sits slouched on a chair behind them. The room in which they are sitting is nearly empty. There are no scalpels here, no microscopes lit with globes of brine, no freshly splenectomied dogs wandering through the room licking fingers. There is no stink of dissected sheep, no sting from turpentine. There is no smell at all. One of the rolling research assistants, Andrew Engell, presses a button and speaks into a microphone. "Everything okay in there?" he asks.

The man with his head in the ring answers yes, and Engell clicks his mouse a few times. A jackhammer blast comes out of the ring and roars out of the speaker next to him.

The man on the slab, his ears plugged with tubes, cannot hear the blare. He cannot feel a magnetic field slice through his head and make some of the atoms in his brain wiggle from side to side. After a pause, the jackhammering starts again as the scanner cuts another slice and repeats the sequence until it has passed from the crown of the man's head to the base of his skull. When the jackhammering stops for good, Engell punches numbers into one of the computers. An image unfurls like a banner rolling down his screen, the data collected in the ring having been turned from numbers into cool shades of gray that could have come from a carefully restored silent movie. It is a picture of the young man's brain, still cradled in his skull.

The picture looks as if someone had opened the man's head between the eyes with a single chop of an axe. The image is no less beautiful than the portraits Christopher Wren drew three hundred and forty years ago. But the image-making is not over yet. Now the young man's brain will be painted in thought. Inside the ring, just over the young man's head, is a tilted mirror. A video projector outside the ring is switched on and reflects a beam of words into his eyes.

> *You are a doctor. You have five patients, each of whom is about to die due to a failing organ of some kind. You have another patient who is healthy.*
>
> *The only way that you can save the lives of the first five patients is to transplant five of this young man's organs (against his will) into the bodies of the other five patients. If you do this, the young man will die, but the other five patients will live.*
>
> *Is it appropriate for you to perform this transplant in order to save five of your patients?*

The young man mulls the choice, the suffering of the one and of the many, the balance between right and wrong, and answers with a click of a button. The white ring jackhammers away, snatching more pictures of his brain and picking up echoes of his thoughts. He answers a few dozen more questions, some innocuous, some tricky, and some agonizing. It is the last category that most interests the copper-haired man slouched in the chair. His name is Joshua Greene. When one of the questions appears on the screen, Greene sometimes speaks up. "Oh, this is the one where you destroy a priceless sculpture to save someone's life. This is the one where you kill your dad for your inheritance."

Joshua Greene is a philosopher. With his disturbing questions and his neurological experiments, he may not seem much like Immanuel Kant or Bertrand Russell. All doubters are welcome to read his massive dissertation on meta-ethics. But Dr. Greene is not content to argue fine semantics or carry out untestable thought experiments. He wants to understand the nature of moral judgments. And so he is in the basement of Green Hall, looking inside brains.

"Some people in these experiments think we're putting their soul under microscope," says Greene, "and in a sense, that's what we're doing. This is what your soul is, if anything is."

Greene is an intellectual grandchild of Thomas Willis—one of thousands. Few of them know of Willis ("You mean Willis, like the Circle of Willis?"), but the fact is that he created the tradition in which they now work. He established a new science of the soul, one that combined anatomical study of the human brain with comparisons to animal brains, experiments, and medical observations. Willis's work formed the four pillars on which the weight of neuroscience rests today.

The first pillar was his discovery that animal spirits traveled through pathways in the brain and that the chemical changes they underwent governed everything in our lives, from emotions and perceptions to walking and sleeping. Today, neuroscientists know that the so-called animal spirits are actually electrical impulses and that they help pass signals between cells known as neurons. Willis's second pillar was his claim that the spirits carried out different functions of the soul as they moved through different parts of the brain. Modern neuroscience has confirmed that the brain is indeed divided into specialized networks of neurons tailored for narrow jobs, such as registering the edges of shapes or the tinge of fear in a voice. Willis made these discoveries about the brain thanks in large part to his third great achievement: demonstrating how similar the human brain is to those of other animals—which neuroscientists now know is a sign of our kinship with them. All animal brains, our own included, have been sculpted by evolution, and our extraordinary human gifts are tinkerings with ancient plans.

In all this work, Willis was guided by his mission as a doctor, hoping to cure the diseases of the soul. This goal was the fourth pillar of his neurology. He believed that all disorders of the brain—neurological and psychological—could be cured by manipulating the atoms that composed it. Willis's steel syrups and crushed millipedes didn't help his soul-sick patients, and for the following three centuries psychopharmacology remained more dream than reality. Today, however, the world is awash in brain drugs—drugs to control epileptic seizures, to keep people awake for days, to send them to sleep, to lift their mood and sharpen their cognition. And those are just the legal ones. Brain drugs have become an enormous business, but the story of their success isn't a simple one. Many compounds have reputations as miracle cures, but their actual effects are modest at best. And when they do work, neuroscientists often can't explain what they are doing to the brain. Much of their popularity stems from our neurocentric culture, which Thomas Willis helped create. A pill, we hope, can cure the diseases of the soul.

In all these ways, Willis had a profound influence over the way we think about ourselves. But he also embedded a paradox at the core of the modern science of the soul. He believed that all his work demonstrated twin souls in man. The sensitive soul was material and thus subject to diseases like any other part of the body, but the rational soul, lodged deep in the brain, was both immaterial and immortal. In other words, it lay beyond Willis's scalpel. The closer you look at Willis's own work, however, the fuzzier the boundary between the two souls becomes. Although the rational soul was supposed to make us uniquely human, Willis could map a vast territory of our inner lives without ever mentioning it. He even claimed that some mental diseases—particularly stupidity and foolishness—actually affected the rational soul. He never explained how something immaterial could be harmed by a mere disorder of atoms.

This paradox still confounds the way we think about ourselves. A man overwhelmed by depression goes to his doctor, who tells him that the depression is a physical disorder, one that he can see for himself on a scan of his brain. He feels relieved because he can feel separate from his disease. The depression is not his fault; it is no more a part of him than a kidney infection, and an antidepressant pill can bring him back to himself.

But everything that neuroscientists have learned in recent years goes against his notion of the self. It is not distinct from the brain, not an unchangeable entity shielded from the assaults of biochemistry. Using the same methods that Thomas Willis pioneered, neuroscientists today are dissecting the self, consciousness, reasoning, language—practically everything that the rational soul was supposed to do or be. It is a testament to Willis's work that he didn't realize how revolutionary his new science would turn out to be or what a quandary he would leave us in today.

—

In his travels through the nervous system, Thomas Willis encountered one of his biggest surprises in the belly of a boar. Most natural philosophers in the seventeenth century believed that some "genital humour" flowed from the male brain to the testicles. The soul had to have some way to reach the boar's semen so that it could ultimately impress itself on the womb of a sow and create a new life. But no one could actually show the connection.

Willis searched for a link but found none. Still, he believed the testicles must have something in common with the brain, because spirit was the only substance in nature that could give such shape and purpose to matter. He was struck by the fact that both the brain and the testicles were surrounded by a filigree of blood vessels. He declared that in both cases the vessels were designed "to carry the most pure flower of the Blood, as it were through the winding Channels of an Alembick, distilled by a long passage." In these vessels, both brain and testicles isolated the blood's most lucid particles of spirit. In the brain, spirits sustained the sensitive soul, allowing it to feel and think and keep the body alive. The testicles simply stored their spirits away. "The Epitome of the whole soul is placed apart for the conservation of the whole species," Willis wrote. This "band of spirits" could enter a womb and give form to a new animal created from a gathering swarm of particles. Because spirits were in limited supply, wise fathers ran the risk of giving birth to stupid children: their brains used up spirits that would otherwise go into their semen.

Today, scientists know that Willis was right to believe that a special sort of spirit flows through both the brain and the testicles. It courses across all life as well and has been in transit for billions of years. It is information.

Much of this flow of information takes place within cells. Their name was coined by Robert Hooke, who gazed through his microscope at boxlike shapes in the tissue of plants. Two centuries would pass before biologists realized that all living things are made of cells, each of which is an oily bag of molecules strung together from atoms such as hydrogen, carbon, and iron. The highest order of business for these molecules is processing information. A cell picks up information from its surroundings, which it represents with special signaling molecules that travel through its interior, sometimes even penetrating the nucleus at its core.

Lying within the nucleus is an encyclopedia's worth of information in the form of a tightly spooled double helix of DNA. DNA is surrounded by a retinue of special enzymes, which spring into action in response to certain combinations of signals. The right signals cause the enzymes to unspool the DNA and make a copy of part of its sequence. This copied text can be edited by the enzymes before it is shuttled to biochemical workshops to serve as a guide for building a new molecule. The molecule may stay inside the cell or slip through the cell membrane to find its purpose elsewhere in the body. Depending on the signal, a cell may spew out crystals of bone or virus-killing poisons. It may commit suicide or divide in two.

Testicles are unusual organs, because their sperm can carry information into another body. If a sperm gets into a uterus and finds a receptive egg, the two cells can fuse, combining their DNA together into a new genome. As the fertilized egg grows into an embryo, its cells get certain signals and respond by differentiating into livers, hearts, and brains.

This song of signals has survived from one generation to the next for billions of years, joining together atoms to impart its melody to a new confabulation. The song is sturdy enough to guarantee that linnet eggs reliably hatch linnets rather than crickets or magnolias. But the information the song carries can gradually change. The recipe shifts. Life evolves. Genes mutate, and if their new code helps their owners survive and reproduce, they become more common with time. As genomes evolve, they can split a species in two or transform old body plans into new ones.

Charles Darwin constructed his theory of evolution in the mid-nineteenth century by turning the ideas of Thomas Willis and his contemporaries on their head. Natural philosophers saw the exquisite design of brains as one of the most powerful demonstrations of God's handiwork, a fail-safe antidote to atheism. God must have designed every kind of brain with a purpose in mind—a purpose unique to each species, which was born equipped with a brain perfectly suited to its life. The argument from the brain's design survived for almost two hundred years, until Darwin realized there was a force that could shape life that was neither random nor mere fiat. The similarities of bodies and brains are actually signs of life's history, of the common heritage of man and monkeys.

—

On a hot day in May 1666, Thomas Willis dissected a man who had been killed by lightning. The bolt had thrown its victim out of the boat he had been rowing. When the body was brought back to town, Thomas Willis came to see it along with Richard Lower and the mathematician John Wallis. They picked up the man's hat and put their fists through the hole the lightning had torn. His doublet had been ripped open and his buttons knocked off. Willis and his friends found spots and streaks across the torso where the skin seemed to be seared and hard, "like Leather burnt with the fire," Wallis later wrote.

The following night the virtuosi returned, along with a crowd of onlookers, to cut the man open. "The whole Body was, by night, very much swell'd," Wallis wrote. The stench that rose from the body was unbearable, but they soldiered on, because such an opportunity might not come again in their lives. Wallis even sent news of the extraordinary event to the Royal Society. "There appear'd no sign of contusion," he reported; "the brain full and in good order; the nerves whole and sound, the vessels of the brain pretty full of blood." They opened the man's chest and found that the burns did not reach below the skin. "The Lungs and Heart appear'd all well, and well-colour'd without any disorder," Wallis wrote. The heavens had struck the man dead, and yet the virtuosi could find nothing changed inside the body.

At the time, Willis was trying to understand how spirits traveled through the nerves. He believed they normally traveled like rays of light, setting off controlled explosions in the body, like matches to trails of gunpowder. To some extent, he would turn out to be right. But what actually travels through the nerves and brains is a substance Willis did not understand. It would turn out to be the same substance that had leapt from the sky and mysteriously killed the man he was carving open.

No virtuoso in Willis's day could have discovered that electricity travels through nerves. It was impossible to study electricity at all until the eighteenth century, when natural philosophers learned how to store it in a jar. They then discovered that the jars could deliver a spark powerful enough to make legs and arms flail. The Italian physician Luigi Galvani later found that in a lightning storm, the severed legs of frogs jerked if he touched their nerves with a piece of metal. But even as it became clear that electricity travels through the nerves, many natural philosophers still refused to abandon the animal spirits. By the dawn of the nineteenth century, the evidence in favor of electricity had become too overwhelming to ignore, and the concept of animal spirits collapsed like an old barn.

The nerves that Willis charted through the body are actually bundles of slender cells known as neurons. Signals can enter a neuron through a set of branches known as dendrites, which may be sensitive to heat, to pressure, or some other stimulus. When a dendrite gets the right stimulus, a chemical reaction takes place, creating an electric pulse that travels to the main body of the cell. The neuron transforms the incoming signal into an outgoing one, which races down another arm called an axon, which splits into thousands of branches. When a pulse reaches the end of an axon branch, it dumps chemicals into the narrow gap separating it from the dendrite of another neuron. The chemicals seep into channels on the surface of the next neuron's dendrites, triggering a new impulse that carries the signal farther through the nervous system.

Neurons evolved out of the ancient biochemistry of microbes, and in the earliest animals they probably formed simple nets, as they do in jellyfish today. But around 530 million years ago a new lineage of animals emerged with many more neurons, which were linked together in new ways. Its new design was so successful that it helped trigger an explosion of new species. The ocean was filled with the forerunners of today's lobsters, oysters, and earthworms. Even the forerunners of people—which looked like headless sardines—began swimming around. When Thomas Willis discovered similarities in the nervous systems of lobsters, oysters, earthworms, and people, he was actually looking at the shared legacy of these evolutionary fireworks.

Biologists today carry on Willis's work, studying these invertebrates in order to learn more about ourselves. By studying the marine snail *Aplysia,* for example, they've learned about human memory. The snail has two thousand neurons, some specialized for touch, some for taste, some for transmitting signals that make the snail crawl or bend. Instead of a brain, the snail has a set of neural clumps, known as ganglia, scattered across its body. The circuitry of these ganglia can amplify some incoming signals and squelch others. The snail's animal spirits follow pathways, but not simple linear ones. Instead, they travel a maze of wiring.

Some parts of the snail's circuit are fixed. If a biologist shocks a snail's tail, a sensory neuron relays a signal from the tail to a ganglion, where the signal makes a motor neuron produce an impulse. The snail contracts its gill covering out of harm's reach. Snails are born with this essential reflex and never lose it. But their nervous system can also shift with experience in order to recognize new dangers. If every shock is preceded by a prod, the snail will learn after a few rounds that the prod predicts the shock. In time, even a touch alone will make the snail recoil.

Here is the science of brutes at its simplest. It is made possible thanks to molecular changes to its nervous system. Once biologists figured out the biochemistry of memories in *Aplysia,* they were able to find the same process—along with many of the same molecules—at work in other animals, from flies to fish to birds to humans. Of course, the memories that we form with these molecules are a lot more complicated than those of a snail, but that is because our own lineage evolved into consummate information processors.

In the earliest days of vertebrate evolution, the tip of the spinal cord swelled into the tangle of neurons known as the brain. Eyes, noses, and other new sensory organs sprouted from it, flooding the brain with information that it could integrate to understand the outside world. Sharks and other fish evolved to smell a blood trail miles away, while others could read the electric fields of animals. Around 360 million years ago, one lineage of fish adapted to life at the water's edge and then to life on dry land. It evolved into a new neurological pageant, into owls that could dive from the night sky at the faintest sight of a mouse, into snakes that could sense heat wafting from a rabbit, into bats that could paint moths with echoed sound. All this diversity grew on top of an underlying plan that has hardly changed over the course of 500 million years. Thanks to the common ancestry of humans and other vertebrates, Willis could see the anatomy of the human brain replicated in sheep, dogs, and fish.

What Willis saw in those animal brains helped him build a new theory for how the human brain works, one that has proven remarkably accurate. And even when Willis made errors—and he certainly made his share of them—they pushed those who came after him closer to the truth.

Willis, for example, offered the first neurological explanation of the reflex. When a sleeping man scratches his face, spirits traveling from his senses into the brain are reflected down to the body without coming to the attention of the rational soul. That claim prompted some natural philosophers to wonder whether the spine might also be able to trigger a reflex. Willis and Lower found no evidence for this, but a century later the Scottish physician Robert

Whytt discovered they were wrong. Whytt cut the heads off frogs and then dabbed their backs with acid. Even without a brain, the frogs could still lift their hind legs and try to scrape the acid away.

The spinal cord is not just "the King's highway," as Willis called it, but more like an extension of the brain itself. The dab of acid Whytt put on the backs of frogs made sensory neurons fire a signal to the spinal cord, where the signal split in two. One signal traveled up to the brain (if the frog still had one) while the other stayed behind, moving among the neurons in the spinal cord. This second signal was passed to motor neurons that control the leg, causing its muscles to contract and relax to make a scraping motion. We experience the same reaction when we kick our leg in response to a doctor's thwack on the knee, or when we instinctively shift our weight from foot to foot in a rocking subway car.

While some signals do not rise above the spinal cord, others take the higher road to the brain. Willis tracked the routes they took, following the brain's streaks and grooves. He could even split their pathways apart with a blunt knife. In the nineteenth century, scientists discovered that these pathways, like the rest of the nervous system, consist of neurons. Staining pieces of brain with silver revealed their delicate axons and dendrites. Neuroscientists could follow the flow of information from axon to dendrite deep into the brain.

Thomas Willis tried to map these pathways in the hope of discovering what each part of the brain did. The cerebellum, he argued, housed spirits that operated regular, involuntary movements, such as the heartbeat and the lungs. It was in the right position to do the job, being joined intimately to the base of the brain, which sprouted nerves separate from the spine, nerves that wandered the body and embraced the heart, lungs, and other organs. Willis tested the idea by removing the cerebellum from a dog. Its death seemed to confirm his hunch.

Although critics accused Willis of making wild speculations, he would prove to be close to the mark. The heart and lungs depend on a small cluster of neurons an inch away from the cerebellum, in the brain stem. They send out signals that travel down the vagus nerve and keep the heart beating and the lungs breathing. The technology that Willis had at his disposal for his experiments—his knives and needles and silk thread—were too crude to let him distinguish between the two regions of the brain, because the damage he did as he explored the cerebellum also harmed the neighboring parts of the brain stem. When neuroscientists in the nineteenth century revealed Willis's error, it was a defeat on the details but a victory on the big picture. He was right to claim that the different regions of the brain carry out distinct jobs.

Moving past the cerebellum, Willis traveled the brain's pathways until he reached the cerebrum, the two large hemispheres that dominate our skull. He argued that the spirits carried out the higher thoughts—such as imagination and memory—in certain zones of the cerebrum. Even in the early nineteenth century, many philosophers and physicians disputed this claim. To divide the soul, they argued, was to deny it. But by the middle of the century, evidence emerged that the soul was indeed divided—or, in the language of neuroscience, the cerebrum is parceled up into several regions that are dedicated to specific functions. Paul Broca, a French neurologist, used methods that Willis had pioneered to take apart the machinery of language. He studied patients who had suffered brain damage that left them unable to speak. After the patients died, he opened up their heads and found that many of them had suffered damage in the same spot, a small region on the surface of the brain just above the left ear.

Neuroscientists continue to map the brain today. They are tracing the boundaries of neuron clusters and following the projections of neurons from one cluster to the next. When clusters are destroyed in car crashes and strokes, neuroscientists can see how the damage changes the way a person behaves. Some people lose their ability to perceive motion. Other people can no longer recognize their own face. Others no longer seem themselves. To study healthy brains, psychologists can conduct tests that tease apart different ways of thinking and create illusions that expose the brain's perceptual shortcomings. Neuroscientists can even stick electrodes directly into the head of people undergoing brain surgery and record how individual neurons respond to different signals—crackling wildly to colors, for example, but falling silent to shapes.

The results of thousands of experiments like these converge on a provisional road map of the brain. Consider a monkey that has been trained to push a button when it sees a picture of a dog. It takes only 40 milliseconds for the monkey's eye to receive the image and start massaging it, heightening contrast between light and dark to make it easier to identify. From the eye, the signals race to a central switching station deep within the brain, where they are directed to the back of its head within 60 milliseconds. Here the signal is spliced and projected onto thirty different visual fields, each specialized for picking out certain features. Some are sensitive to edges, some to shading, some to corners. A wave of signals surges from these visual fields across the monkey's cortex, moving forward toward its ears. In this region, known as the temporal cortex, parts of the brain respond to more complicated features, such as shapes and movement. By 100 milliseconds, the signals have moved from the temporal cortex to the front of the monkey's brain, just behind its forehead. This region, known as the prefrontal cortex, sorts through its rules for action—such as *see dog, press button.* Now the flow of information starts moving back toward the body. The neurons in the monkey's body-map start firing, setting up commands to move its fingers. Within 160 milliseconds these commands are hurtling down the spinal cord, and the finger presses the button only 180 milliseconds after the picture of the dog first appeared.

The information racing through this monkey's brain takes a simple path compared to paths of a person daydreaming, or recalling her first blizzard, or dividing 234 by 6. And mapping such complicated paths calls for a particularly sophisticated kind of surveying equipment, such as magnetic resonance imaging scanners.

An MRI scanner works by setting up a magnetic field so intense that it forces some of the atoms in the brain to line up together. The scanner then sends out radio waves, which push the atoms a few degrees off kilter. When the waves are turned off, the atoms rotate back into line again, and along the way they give off pulses of energy. The scanner uses these pulses to reconstruct a slice of the brain, and a computer can stack a series of slices together to form a three-dimensional scan of the brain.

If MRI could do no more than this, it would still be a revolutionary technology. It translates each brain into a cloud of coordinates in the mathematical space that Descartes invented, allowing a computer to calculate the difference between brains by measuring the distance between their coordinates. Neuroanatomists can now see the variations in dozens or hundreds of brains in a single image. They can color-code a collective brain to show how much each of its regions is shaped by genes or by experience. Or they can create a map of the brain through time by scanning a person's head as she or he grows up through childhood and adolescence. Thomas Willis had to wait until his patients died before he could get any glimpse of how diseases changed brains. Today neuroanatomists can watch schizophrenia ravage brains years before any noticeable change in behavior, as a spreading wave of destruction travels over time from the back of the brain to the front.

But MRI scanners can do more than this: they can capture snapshots of thought. As neurons receive signals from their neighbors, they devour oxygen, which triggers a rush of blood to replenish the supply. MRI scanners can pick up these microscopic tides, because the oxygen molecules surging toward an active neuron give off a distinct pulse of radio waves. In the 1660s, Willis studied animal spirits by following the brain's blood with injections and dyes. Today, neuroscientists follow the blood to trace thought itself.

The images of thought created with MRI over the past decade clearly show that the brain contains many modules dedicated to different kinds of thought. Some, like Broca's area, help with language. Some are more active when people read nouns instead of verbs. Some are more active when a bilingual person reads Mandarin instead of English. Even when people listen to different kinds of jokes, they produce different patterns in an MRI scan. A joke that depends on the meaning of words makes the semantics-processing part of the brain work hard, while a pun makes the region that processes sounds into words light up.

While the brain may be a collection of modules, none of them can work alone. During any mental task, information reverberates in a far-flung network of regions that light up like constellations on an MRI scan. These networks are constantly changing over scales of seconds, minutes, and decades. Connections grow strong or weak; old nodes disappear and new ones take over.

This flexibility is crucial, because fueling the brain costs us dearly. The activity of a single neuron uses up so much energy that less than 1 percent of the neurons in the cortex can be active at any moment. With such a limited budget of energy, the brain simply cannot take in all the information available to its senses. It must make up for its shortcomings with elegant strategies for picking out only what matters. But what is important one second may become unimportant the next, and so the brain also needs to continually rearrange its networks, refocusing its attention on perceptions that do the best job of predicting the future. A driver may focus his attention on a driveway because he knows that his neighbor has a habit of lurching out into traffic. His brain makes the neurons in that part of his field of vision more sensitive and boosts their signal by as much as 30 percent. Most of the time the brain refocuses itself automatically without our awareness. Willis's sensitive soul is at work, altering the pathways that the animal spirits travel through the brain.

Willis believed that these pathways were altered not through the influence of humors or the stars, but by chemicals. Today neuroscientists are figuring out exactly what those chemicals are. One of the most important is dopamine, which is produced by a few thousand neurons buried deep in the brain stem. When an animal unexpectedly finds a reward—be it food, water, or sex—these neurons release a surge of dopamine from their thousands of branches. Suddenly many parts of the brain become focused on the reward and how the animal can find it again. With more practice, an animal begins to associate certain cues with the reward. Before long the dopamine is triggered by the cue, rather than the reward. It gives the animal an invigorating feeling of anticipation, a notion of cause and effect.

Like other animals, we humans rely on dopamine to do anything new: to learn how to walk, to shoot a basket, to get a can of iced tea from vending machines in Tokyo even when we can't read Japanese. And as we gain more expertise, dopamine gives us its feeling of exhilaration and happiness not with the reward itself but earlier, with the things we know presage the reward. A gambler shouts for joy not when he is cashing in his chips, but at the craps table, when the dice turn up seven.

Most of the branches of dopamine-producing neurons reach into the prefrontal cortex, the outermost layer of the front third of the human brain. This region is crucial for figuring out cause and effect and using that information to reach a goal. When the prefrontal cortex is damaged, the fabric of reality begins to unravel. People with prefrontal cortex damage may still be able to add cream to their coffee and stir it with a spoon; they may just stir first and pour later.

Simple games can reveal how the prefrontal cortex brings order to our lives. One involves dealing a deck of cards decorated with diamonds and circles and other shapes. Volunteers look at the cards and match pairs of them without being told what the rule of the game is. The rule might be to match cards according to color. The scientists tell the volunteers if their match is right or wrong, and they try again. It doesn't take long for most people to figure out the rule, and brain scans show that they use their prefrontal cortex to do it. Once volunteers figure out what the rule is, the prefrontal cortex calms down. If the scientists change the rule without telling the volunteers—perhaps now cards have to be matched by shape—the volunteers suddenly find themselves making wrong choices. Their prefrontal cortex switches on again.

Earl Miller of the Massachusetts Institute of Technology and Jonathan Cohen of Princeton University have used this game and others like it to put together a particularly persuasive model of how the prefrontal cortex works. As a card-sorting game begins, information starts to flow from the senses to the prefrontal cortex. The neurons in the prefrontal cortex influence the choice a volunteer makes. Pairing cards correctly triggers a surge of dopamine, which strengthens the connections between the neurons. With each trial, more neurons in the prefrontal cortex join into the card-sorting network, turning the simple association between pairs of cards into a general rule. As this rule proves reliable, a correct guess no longer brings a jolt of dopamine. The prefrontal cortex becomes quiet, and the brain now automatically follows the newly constructed pathway from cue to response.

When the rules switch and volunteers start getting wrong answers, a special part of the brain detects the conflict. A band of neurons in the cleft of the brain's hemispheres becomes active. Known as the anterior cingulate cortex, it sends out signals to the prefrontal cortex to call it back into service again. The prefrontal neurons learn the new rules and send signals to the rest of the brain that reorganize it from its old response to a new one. The signals might downplay the importance of colors while highlighting shapes instead. By favoring certain incoming signals, the prefrontal cortex shuts out distractions that interfere with learning a new rule. Once the brain's new rules start producing good results and there is no conflict anymore, the anterior cingulate cortex quiets down again.

Rules for a game of cards can be easily learned and unlearned. But rules we follow for years aren't so easy to overcome. Americans in England often look the wrong way as they step off a curb, even if they know full well that cars in England drive on the left side of the road. A psychological experiment known as the Stroop test can reveal the nature of this mistake. People are asked to look at a word printed in colored letters and name the color they see. The word "red" in green letters slows down people's response enormously, because the two competing responses vie to be our choice. Brain scans show that a Stroop test summons the anterior cingulate cortex into action. As it senses a conflict between responses, it recruits the prefrontal cortex. Boosting the weaker response of naming the color over the habitual response of naming the word, the prefrontal cortex lets a person get the answer right. The same thing happens when we struggle with a name on the tip of our tongue, as a swarm of similar names competes to emerge from our memory. The same network even becomes active when people lie, because a lying brain has to struggle between the strong habit of telling the truth and its new goal of deception.

Miller and Cohen's model does a good job of explaining how the brain makes up its mind, but it doesn't capture the full story. Dopamine is only a simple signal, a flag raised at an unexpected reward. It can work only if the brain has already set up its scale of rewards, so that it can decide that some things are more valuable than others. Why should rolling seven at the craps table feel rewarding, while the sight of a starving child doesn't?

Emotions are the reason why. Human emotions are descended from ancient programs that guide animals away from things that might harm them and toward the things they need to survive. The chemicals produced in our bodies at the sight of an oncoming truck aren't profoundly different from the ones produced in a mouse by the sight of an oncoming cat. In both human and mouse, a surge of hormones speeds up the heart and creates an urge to flee or hide. If a mammal is frustrated in a search for sex or territory, it may become enraged. If it is separated from its family, it may feel anxiety.

In the Renaissance many philosophers and physicians believed that the mind's emotions made themselves felt in the body by mystical sympathies. But Thomas Willis reduced the connection to a network of nerves—specifically, the nerves that sprout from the brain above the spinal cord and send their branches to the face, heart, lungs, bowels, and groin. Today, these nerves still carry a name redolent of Renaissance mysticism: the sympathetic and parasympathetic divisions of the nervous system. But thanks to Willis, cosmic sympathies have given way to the mechanical motion of spirits carrying emotional signals from the brain to the body.

Today, neuroscientists are mapping the pathways of the emotions within the brain itself. Fear, for example, depends on the amygdala, an almond-shaped bundle buried deep in the temporal lobe. The amygdala encodes the primal fears we are born with and cements our association with new terrors. Fear strikes us suddenly because the amygdala doesn't have to wait for the higher regions of the brain to work over sensory information or run it through some abstract set of rules. Neuroscientists have been able to activate the amygdala by flashing pictures of angry faces at people for only forty milliseconds—too fast for them to become consciously aware of them. In that brief time, the amygdala may be able to take a rough measure of a situation and detect anything that looks or sounds particularly dangerous. It then sends out a signal that makes hormones race through the body to prepare it to react. In other words, the amygdala acts almost like a little brain unto itself.

As the prefrontal cortex evolved in our ancestors, it became intimately wired into these primitive circuits, transforming simple emotions into nuanced feelings. One region, known as the orbitofrontal cortex, came to play a particularly crucial role. It takes in signals from many emotion-linked regions of the brain and then crunches them like a hedge fund manager, making calculations about the relative value of things. It makes us savor the taste of chocolate when we're hungry and recoil in disgust when we've had too much to eat. It puts emotional value on abstract things such as money by associating with them all the things they signify. While other parts of the prefrontal cortex handle the "how" and "what" questions in life, the orbitofrontal cortex appears to handle the "why."

Discoveries like these show how foolish it is to try to dig a deep trench between emotions and rational thought. Emotions sharpen our senses, focus our brains, and help us remember things more clearly. The prefrontal cortex returns the favor by moderating the emotions. Although people can sometimes consciously tame their emotions, this sort of regulation probably goes on all the time without any conscious thought whatsoever. Recall the Stroop test, with its confusing colored words. Another way to slow people down on a Stroop test is to ask them to name the color of upsetting words, such as "murder" or "rape." In those pauses, an unconscious conflict is being negotiated, as regions such as the amygdala focus on the emotional charge of the words and interfere with identifying their color. The brain's conflict monitor, the anterior cingulate cortex, swings into action, recruiting the prefrontal cortex to downplay the emotions so that people can give the right answer.

Some common mental disorders may emerge when parts of this emotional circuitry corrodes. The sight of dirty hands makes us want to wash them, but the urge normally subsides after a few moments under the faucet. People with obsessive-compulsive disorder may have a defect in their emotion-regulation network, so that the urge can't be downgraded, and they find themselves scrubbing their hands dozens of times a day.

When depressed people think about emotionally charged things—even individual words—their emotional circuit reacts much more strongly to sad cues than to neutral or happy ones. They are overwhelmed by negative thoughts, perhaps because their orbitofrontal cortex can't give those thoughts the proper weight. In other words, the distorted, darkened spirits of Willis's melancholy live on. They have become the distorted, darkened information-processing pathways of the depressed brain.

—

Thomas Willis wanted not only to understand melancholy but to cure it. European physicians had been treating melancholy for centuries, but Willis represented a radical break from their traditions. Even in the 1630s, the English physician Richard Napier was still treating melancholy by drawing on astrology, Renaissance magic, Galen's four humors, and Christian prayer. Only thirty years later, Willis relied on none of these, searching for remedies instead in the chemistry of atoms and the physics of mechanical matter. To cure melancholy, he prescribed his famous steel syrup, promising that it would wash out the corpuscles of salt and sulfur from the blood and "shut the little mouths of the gaping vessels of the brain." Willis also prescribed pleasant talk, jesting, and other diversions for melancholy, but for him, talking and jesting had a physical effect on the brain, just like steel syrup. They pulled the melancholy spirits away from their distorted paths and back to their normal orbits.

Most doctors who came after Willis followed his lead. In the eighteenth century, they thought they could change people's temperaments by tightening slack nerves. In the nineteenth century, chemists searched for plant compounds—aspirin from willow bark, hyoscine from henbane, digitalis from foxglove, cocaine from coca leaves, morphine from poppies—that altered the nervous system.

But countertraditions also sprang up to challenge this image of the brain. Romantic poets such as William Blake condemned the sterility of modern science, sharing Anne Conway's dream of the soul liberated from the tyranny of dead matter. In the early nineteenth century, Lady Conway's fellow Quakers built retreats for the insane, where they treated their patients without drugs or physical restraints. They cured the insane with "judicial kindness," talking with them to help them return to their inner light. Psychotherapy, as this relationship between doctor and patient became known, grew more and more popular as the nineteenth century passed. Drugs could stop some ills of the brain—morphine could kill pain, aspirin could stop a headache—but hysteria, depression, and psychosis remained untreatable. To many, psychotherapy seemed a more effective and humane treatment. And just before the close of the nineteenth century, it gained its most important follower: a Viennese neurologist named Sigmund Freud.

Freud had started out as the sort of neurologist who would have made Thomas Willis proud. As a young man, he studied the nerves of crayfish and lampreys, marveling at their underlying similarities to human neurons. Freud even set out to create what he called a scientific psychology based on the mechanical workings of the brain and nerves. He pictured an energy passing from one nerve to the next, following Newton's laws of motion. But it quickly became clear to him that neurologists knew precious little about the workings of neurons and even less about how their workings produced psychological experiences. He abandoned his scientific psychology as a futile dream.

Instead, Freud decided to build an epic of the soul. He gave it a cast of characters, the most extraordinary being the id, or unconscious. Freud argued that energy from the body's basic drives entered the brain through the id, which strove to achieve goals such as food or sex without our conscious awareness. If those drives became blocked—perhaps as the mind tried to repress a traumatic memory—the trapped energy would become poisonous and create hysteria or some other mental disorder. Freud's mission became helping his patients discover their own repressed desires and traumas and, in recognizing them, cure their own minds.

Psychoanalysis fascinated the twentieth century. By the 1950s, psychoanalysts were running most of the psychiatry departments at American universities. Their language invaded movies, magazines, and television. Sometimes they spoke as if they had the power to change the fate of nations, promising that there would be no more wars if only the world's leaders went into therapy.

But psychoanalysis eventually revealed some serious shortcomings. Psychoanalysts continued to cling to the notion of nerve energy even after it had been proved wrong. They shied away from scientifically validating their ideas, so that each analyst's intuition became a final authority. Over time, they created baroque explanations for mental disorders. In 1937, a psychoanalyst claimed to have cracked the mystery of obsessive-compulsive disorder: the ego had "already begun to adopt protective measures at the time of the original anal sadistic level of libido organization, so that the patient never reached the phallic Oedipus complex."

By the 1970s psychoanalysis was in decline, and it has now fallen so far out of favor that most psychiatric residents no longer get any training in it at all. But hubris was not the only reason for its slide. It was also under attack from Thomas Willis's old tradition, which had found its strength again. A doctor faced with obsessive-compulsive patients today is less likely to help them get beyond the anal sadistic level of libido organization than to tell them to swallow a pill.

The modern resurgence of brain drugs began just after World War II. Thorazine and other new compounds helped the mentally ill like nothing that had come before. A barber who had been sunk in a stupor for years took Thorazine and promptly woke up, declaring that he wanted to get back to cutting hair. Another patient responded to medication by asking for billiard balls and began to juggle. Neuroscientists discovered that these new drugs were able to change the levels of dopamine and other neurotransmitters in the brain. It suddenly seemed as if the cure for madness was simply a question of adjusting the brain's alembic, just as Willis had promised.

Despite the lab-coated giddiness of the 1950s, psychopharmacology didn't change the world overnight. Thorazine and drugs like it proved to have awful side effects, and the search for better drugs slowed down as governments began demanding years of testing. Only in the 1970s did pharmaceutical companies start selling new drugs that could reduce symptoms as well as the old drugs had, with fewer harmful side effects. Now, at the start of the twenty-first century, the human brain is glutted with Prozac, Paxil, Zoloft, Xanax, and Ritalin. Treating the mentally ill helped make Willis a rich man, but he never dreamed of the wealth that would be reaped from chemical remedies today. Antidepressants alone bring in twelve billion dollars a year in the United States.

There's no doubt that some people have benefited from this new generation of drugs. When people find relief from depression with Prozac, an MRI scan shows that their brains change: they look more like the brains of healthy people. Just as Willis had hoped, drugs can indeed alter the paths of spirits. But none of these drugs has proven to be a magic bullet. After six to eight weeks on Prozac or other leading antidepressants, only 35 to 45 percent of people with major depressions really begin to feel well again. The rest improve only a little or not at all. Even when these drugs work, scientists are still debating one another about what they are doing to the brain. It's even possible that for many people the chemicals in these drugs are not the source of their power. Some studies suggest that most of the effectiveness documented in antidepressants is due to the placebo effect. These drugs may often work, in other words, because of the faith they inspire.

Thomas Willis was one of the original architects of that faith. He isolated the soul from stars and demons and made the chemical workings of the brain the key to sanity and happiness. Just as important, he helped make the brain a familiar thing. Christopher Wren's drawing of the brain became a map of the soul's entire world. People came to recognize it and to agree on what they saw. In 1762, for example, the British painter William Hogarth made the brain part of a nasty allegorical attack on the Methodist Church in an engraving entitled "Credulity, Superstition, and Fanaticism." A Methodist minister preaches to a rabble, his wig tipping off the back of his head to reveal a Catholic tonsure. From one of his outstretched hands dangles a witch on a stick, and from the other a little demon. The congregation—a sea of apes—convulses with an enthusiasm that looks like madness. A lewd minister preys on an ecstatic girl in the congregation, slipping an icon down her bosom. A Jew worshiping in the church kills lice between his fingers, with a knife resting on his open Bible. In the foreground of the church a woman gives birth to rabbits and a boy vomits up nails. A century before, the Royal Society might have dispatched virtuosi to record the details of these wonders. But in the Enlightenment, they had become emblems of the fanatic, unreasoning mind.

To drive home his point, Hogarth placed a brain in the lower right-hand corner of the picture. Almost no one in his audience would have seen an actual human brain, but Hogarth felt no need to identify it with a label. The brain simply sits there in the church, with the frontal poles of its cerebral cortex pointed upward, exposing the Circle of Willis and the nerves sprouting from the brain's underside. It is identical to Wren's drawing of a century earlier, its position intended to reveal God's handiwork and to place the human brain in contemplation of heaven above.

But Hogarth's brain also sprouts a thermometer from its frontal poles—another icon of the scientific revolution, representing objective measurement over subjective judgment. Its scale does not measure degrees of heat, but degrees of madness, running from the melancholy lows of suicide and despair to the manic highs of lust, ecstasy, and convulsions.

Today the brain is even more recognizable, even more steeped in meaning—although that meaning has morphed. A healthy brain represents not a soul in proper Anglican form, but a well-regulated psyche, a chemically fulfilled self. Reason, once dedicated to salvation, is now the measure of IQ tests, admission to colleges, pay scales. And the highs and lows of Hogarth's thermometer have been replaced by the fiery reds and chilly blues that neuroscientists use to color the MRI scans of abnormal brains.

In brain scans we see our psychic woes and their cures. Far from badges of shame, the scans are embraced by many people with mental disorders. Patients-rights groups put brain scans in their newspaper ads, pointing to scans as proof that mental disorders are not a matter of moral failure but just a question of bad chemistry. We can see for ourselves, it seems, that mental disorders have a physical home in the physical brain. The psychiatrist Wayne Drevets, an expert on depression at the U.S. National Institute of Mental Health, once described a patient who had an easier time dealing with breast cancer than with depression. "With depression there was nothing tangible to point to or explain to people, even herself," he said. A brain scan gave her something tangible to look at—a shrinking brain volume, perhaps, or a surfeit of blood flowing into the amygdala. They show up on a scan as clearly as a tumor in a mammogram.

A picture of depression is comforting because it seems to separate the viewer from the disease. A drought of dopamine in your brain has as little to do with your true self as a gallstone. You can get rid of this sort of physical disorder with a drug, which simply erases those glowing highs and lows from your brain scans. Pharmaceutical companies certainly recognize this belief, and they play on it in their advertising. When GlaxoSmithKline launched an ad campaign in 2002 for the antidepressant drug PaxilCR, their slogan was, "I'm back to being me." Me, in other words, is something distinct from the vagaries of the brain itself.

If that were true, these psychiatric drugs wouldn't alter a person who is healthy and free of mental disorders. But that's not the case, as a recent experiment showed. A group of healthy people who were given antidepressants for a few weeks became friendlier and more socially dominant. When were they their real selves—before the drug, or after? Perhaps, if the self is actually encoded in the brain's synapses, the answer is both. Perhaps the gulf between a brain scan and the person looking at it isn't all that wide.

This paradox is not new. We are still wrestling with the contradictions of Thomas Willis's neurology. Willis believed that the sensitive soul was a material system that encompassed the brain, nerves, and spirits, and that it coexisted with a rational soul that was both immortal and immaterial. Yet he was such a good neurologist that he ended up betraying his own claims. If we do have an immaterial soul, scientists today have no hope of finding it, because that which does not obey the laws of nature is beyond science's scope. And yet just about everything that Willis claimed that the rational soul does has fallen within its scope. The human brain uses distinctive networks of neurons to carry out them out, networks no different in kind from the sort that carry out the business of Willis's sensitive soul.

Reasoning, for example, leaves a mark on an MRI scan. Actually, it leaves many marks, because particular networks in the brain specialize in particular kinds of reasoning, such as deduction, induction, and analogy. These circuits did not appear in our brains out of nowhere. We can see their evolutionary precursors in other primates, who can discover abstract rules with their prefrontal cortex, just as humans do. Mathematics has an evolutionary heritage as well. The network we humans use to solve math problems includes regions of the brain that have other uses—for example, a region that also processes the meaning of words. But the network also encompasses one special strip of the cortex just over the left ear. This math zone is designed for creating an abstract "number line" on which we array numbers that we compute in our heads. Monkeys have rudimentary math skills—they can tell the difference between eight apples and nine, for instance—and they use a smaller version of our number line.

The anatomies of reasoning and mathematics have been easier for neuroscientists to map than another faculty of the rational soul: its consciousness of itself. Progress has been slow in part because the words "consciousness" and "self" have a way of slipping around in the semantic mud. But neuroscientists are now taking little steps forward with some basic experiments. One of the most promising theories to come out of these experiments is that consciousness consists of a brain-wide synchrony. When we become aware that we are seeing or feeling something, a lot of neurons start producing synchronized pulses together between thirty and fifty times a second. It's possible that this synchronization joins together many parts of the brain at once, turning them into a giant global workspace where all our perceptions can assemble into a conscious whole.

Finding the mechanisms of consciousness will not mean that we lack a true self. It's just that this self looks less and less like what most of us picture in our heads—an autonomous, unchanging being that has a will all its own, that is the sole, conscious source of our actions, and that distinguishes humans from animals. All animals probably create some kind of representation of their bodies in their brains, and humans simply create a particularly complicated model. We infuse it with memories, embellish it with autobiographies, and project it into the future as we ponder our hopes and goals.

The human self did not reach this complicated state on its own. Thought is more like a node in the social network of our species. All primates are remarkably social creatures, and our ancestors ten million years ago were no different, depending on one another to escape leopards and to fight other primate bands for fruit trees. Under these conditions, our ancestors evolved into political animals, capable of creating coalitions and settling conflicts. They squabbled over food, competed for sex, and made their way up and down the social hierarchy. By five million years ago our ancestors had become upright walkers who probably traveled in bands a few dozen strong. They evolved an ability to understand the minds of other people and to predict what other people would do. They found happiness in cooperation and trust, which helped them search for food and shelter together.

The result of this evolution was an awesome social computer. The human brain can make a series of unconscious judgments about people—recognizing their faces, judging their emotions, and analyzing their movements—in a fraction of a second. In recent years, neuroscientists have been mapping out the networks that make this social intelligence possible, and one of their most astonishing discoveries is that a picture of the brain thinking about others is not all that different from a picture of the brain thinking about oneself. Some neuroscientists think the best explanation for this overlap is that early hominids were able to understand others before they could understand themselves.

As strange as this might sound, it makes evolutionary sense. There could have been a huge advantage to a hominid in understanding the intentions, feelings, and knowledge in the brains of others. Only later did a full-blown human self emerge from the same neural circuitry, like a mental parasite. This theory might help explain the way our brains sometimes blur the line between ourselves and others. Our overlapping circuitry may make some people prone to projecting what they can't reconcile with themselves onto someone else. Our own thoughts become communications from aliens transmitted through the fillings in our teeth. A ghost nudges the pointer on a Ouji board. A divining rod dips.

The self, neuroscientists are finding, has an ancestry, a physical wiring, and biological weaknesses. So do consciousness, reasoning, mathematics, and the other faculties that Thomas Willis believed were the business of the rational soul. And the same holds true for what Thomas Willis believed was the highest calling of the soul and the ultimate purpose of all his anatomizing—understanding what is good and bad.

—

For Thomas Willis, morality was a straightforward matter. God endowed man with a rational soul, which determines right and wrong through reason. Willis founded the science of neurology on this belief, convinced that only with a healthy brain could a rational soul exercise right reason. The delusions of a fever and the rantings of a false religion were equally dangerous threats to a person's moral judgments. Ultimately, a clouded brain could deprive a soul of salvation.

Thanks to Willis's wayward student John Locke, philosophers in the eighteenth century stopped looking to the physical workings of the brain to understand morality. An Enlightenment philosopher looked instead in the realm of ideas and reason. Immanuel Kant argued that reason alone showed that morality boiled down to a few rules: that we must not use other people purely as a means to our own ends, and that one should personally follow a maxim only if it could be turned into a universal law. In later years, other philosophers, such as John Stuart Mill, found another explanation for right and wrong: they are measures of the happiness brought to the greatest number of people. While Mill and Kant might disagree about the foundations of morality, they agreed on one thing: we make moral judgments by reasoning about right and wrong, which are part of the real world that lies outside the mind—a school of thought known as moral realism.

In recent years, a growing number of philosophers have become skeptical about moral realism. No matter how moral realists try to prove the objective reality of moral judgments, sooner or later they all end up sounding like the parents of little children, driven to saying, "Just because!" Why is setting a cat on fire wrong? Because it causes unnecessary suffering. Why is unnecessary suffering wrong? Because a person who is fully informed and fully rational would say that it is wrong. Why would such a person say it is wrong? Just because!

The Enlightenment philosopher David Hume was the first to declare that we do not approve of good acts because we rationally recognize them as good, but because they just feel good. Likewise, we call things wrong because we have a feeling of disgust for them. Moral knowledge, Hume wrote, comes from an "immediate feeling and finer internal sense," not by a "chain of argument and induction."

Hume's ideas were promptly buried in Kant's avalanche of reasoned morality and would not be dug up again for a century, when Charles Darwin realized that evolution shaped not only bodies but thought as well. If philosophers really wanted to answer many of their biggest questions, they should get back to natural philosophy, to Thomas Willis's approach to the brain. "Origin of man now proved," Darwin wrote in his notebook in 1838. "Metaphysics must flourish. He who understands baboon would do more toward metaphysics than Locke."

Inspired by Hume and Darwin, today's opponents of moral realism have created a new theory of moral judgment. Calling themselves social intuitionists, they argue that when people decide what is right or wrong, reasoning plays a minor role. Most of the time, moral judgments occur in the hidden world of unconscious emotional intuitions. These intuitions have a long evolutionary history in our primate ancestors. Groups of chimpanzees, for example, will punish misbehaving individuals. A zookeeper once witnessed this proto-morality when he began to feed his chimpanzees only after they all had come into an enclosure. Sometimes a few young chimps dallied outside for hours. The other chimps would remember their misdeed, and attack the stragglers the following day.

Chimps may be smart, but they don't read Kant. The stragglers were punished not because the chimps reasoned about their behavior, but because they got angry. According to the social intuition model, similar emotional responses underlie human moral judgments as well.

Social intuitionists don't claim that humans are hardwired with one type of morality. That would be like saying we are all hardwired to speak Hindi. All people are born with an instinct for learning the rules of grammar, but depending on where they grow up, they become fluent in Hindi or English or Farsi or Xhosa. As children are learning languages, they are also picking up the particular morality of their culture. They end up with both a mother tongue and a mother morality. These intuitions make us judge other people in certain ways, and they also influence how we conduct our personal lives. But if the brain's circuitry is damaged, these intuitions may not form, and a child may not develop into a moral adult. There's evidence of this sort of damage in the brains of psychopathic criminals. They fail to respond to the sight of a crying child the way other people do—even nonpsychopathic murderers feel a twinge.

Social intuitionists do not ban reason from moral judgments altogether. We use reason to sort through a complicated dilemma, but it's a slow operation that runs awkwardly compared to our swift intuitions. More often, reasoning brings up the rear, creating after-the-fact justifications for our snap judgments.

Joshua Greene, the resident philosopher in the basement of Green Hall at Princeton, has been able to watch social intuitionism at work in the brain. In 1999 he decided to use Princeton's MRI scanner to find clues to the philosophical question of how people decide what is right and wrong. He set up an experiment in which he scanned the brains of people struggling with difficult moral decisions.

Greene's subjects had to decide whether it was right in different situations to harm someone to save other lives. In some of his questions the harm was personal, in others impersonal. In a typical personal question, he asked his subjects to picture themselves standing on a bridge over a trolley track. A runaway trolley is about to kill five track workers, but you can save them if you push a large man next to you off the bridge, stopping the trolley. One man dies to save five others. Is it appropriate to push the man to his death?

Greene then posed the same question to his subjects impersonally: imagine that the runaway trolley is approaching a fork. The track is set so that it will veer left and kill five workers. But you can switch a button and make it veer right, where it will kill only one. Is it appropriate to switch the tracks? The body count is the same in both, and yet to many people killing someone impersonally feels like a regrettable necessity, while pushing someone feels horribly wrong.

Over the course of an hour, Greene's subjects answered dozens of these sorts of questions, along with morally neutral questions for comparison. After each subject had run through all the questions, Greene then spent hours turning the raw data he had just collected into images of the brain, showing which areas responded more strongly for each kind of question.

Greene predicted that personal and impersonal moral judgments would draw different reactions, because they trigger different circuits in the brain. A scenario that calls for us to kill someone with our bare hands is the sort of moral violation our ancestors could have encountered and recognized millions of years ago. It automatically summons up strong emotional reactions that make us feel that the violation is wrong—regardless of what our reason might tell us about the good that might come out of it. A scenario that calls for us to throw a switch to move a trolley from one set of tracks to another doesn't trigger this emotional response because it is not the sort of crisis our ancestors ever found themselves in. As a result, it triggers no snap moral judgment. Instead, we have to rely more on the abstract reasoning of the prefrontal cortex—weighing costs and benefits—to decide what is right and wrong.

The scans matched his predictions. Impersonal moral choices—ones that did not involve direct, face-to-face harm—made the brain work in much the same way as it did during non-moral choices. Both of them tended to draw blood into the same network in the prefrontal cortex important for cool, logical thinking.

The personal moral choices had more of an effect on three other areas. One of the regions, near the front of the brain, is important in understanding other people's minds. Another region, located just behind the anterior cingulate cortex, flares up in many emotional situations, such as when people read emotionally charged words. And the third region, known as the superior temporal sulcus, is responsible for detecting biological motion; it registers information such as eye movements and lip movements that help us recognize other people.

Each of these regions may have its own part to play in a personal moral judgment. The superior temporal sulcus helps make us aware of others who would be harmed. Mind-reading lets us appreciate their suffering. The emotional region may trigger a negative feeling—an inarticulate sense, for example, that there is something just plain wrong in killing someone.

It seems that for impersonal moral questions, this emotional circuit is only weakly active. Our brains rely more on the prefrontal cortex to apply rules and concepts to the choice. This impersonal network may become active when we face quandaries the modern world presents us, ones that arrive without the cues that would summon up the older, more emotional moral circuits. Instead of making quick, unconscious judgments about right or wrong in these situations, the prefrontal cortex may balance the goods and bads on a mental balance sheet. Sometimes we act like Kant, sometimes like Mill.

These two circuits are not mutually exclusive. It's likely that they work simultaneously, but in many cases one circuit is strong and the other is weak. Sometimes they produce opposite responses with equal strength, and our brain has to struggle between them. Greene has found evidence for this struggle in the time it took for his volunteers to answer his questions. When people decided that a personal moral action was inappropriate—killing a baby you've just given birth to, for example, just because you're scared of the responsibility—they responded with a swift *no*. But they took twice as long to choose to do personal harm—as in the scenario in which your village is hiding from enemy soldiers and your baby starts crying. Do you let the soldiers kill all the villagers, or do you smother your baby?

Greene's subjects hesitate in the same way people hesitate

when they take the Stroop test, in which their brains are pulled by two competing responses to a colored word. In the Stroop test, scientists have found, the anterior cingulate cortex swings into action as a conflict monitor. Green found that contemplating an agonizing dilemma like the wartime baby smothering delayed the response from his subjects. And in these situations, the anterior cingulate cortex became active once again. We struggle with moral judgments, it seems, when these two circuits come up with strong, competing conclusions. It is only then that the circuits for judging right and wrong throw off sparks. It is only then that we get a glimpse of the joints and sinews of our moral anatomy.

—

Thomas Willis was afraid of this. He wrapped his books in disclaimers, in dedications to archbishops, in promises that he was not in league with atheists. He did not want his discoveries about the brain to upset the balance between the immaterial and the material, to lead people away from the path to goodness. Nevertheless, Thomas Willis's work at Beam Hall led straight to the work going on in the basement of Green Hall, where a machine eavesdrops on the moral circuits crackling inside a man's head. This is not a cause for fear, though. The world into which neuroscience is bringing us is not a nightmare of blind atoms.

Instead, neuroscience is making us more aware of how morality is intertwined with our deepest emotions, our quickest intuitions. Sometimes we can trust these intuitions, but other times we may need to push them off. Neuroscience sheds light on how different moralities take shape in different brains, and how those moralities can then crash into each other, causing grief and confusion on all sides. If we step away from the illusions of moral realism, of our mistaken conviction that we perceive right and wrong the same way we perceive red and blue, there might be less grief and confusion in the world. We would do well instead to bear in mind how morality took shape over millions of years, as a profound concern for others.

Thomas Willis's anxiety about his new science sprang from a

false dichotomy, one of many that have dominated Western views of the brain and the self. An equally meaningless one forces us to choose whether our mental disorders are mental or physical. It lurks behind the controversy over whether psychotherapy or drugs have more power to heal the mind. But if the dichotomy were real, it would be impossible to explain the placebo effect: the fact that a sugar pill can sometimes have the same effect on the mind as Prozac—an effect, it should be noted, that can be seen on an MRI scan. If the dichotomy were real, it would be impossible to explain the fact that when psychotherapy and antidepressants ease depression they change the activity of the brain in much the same way.

Our souls are material and yet immaterial: a product of chemistry but also a pulsing network of information—a network that reaches beyond the individual brain to other brains, linked by words, glances, gestures, and other equally immaterial signals, which can leave a physical mark as indelible on a scan as a stroke or a swig of barium, and yet never become merely physical themselves.

The science that Thomas Willis pioneered has brought us to this remarkable place, but now Willis and his methods are no longer enough. It's a good time in the history of science to recall the visits Willis paid in the mid-1660s to his most extraordinary patient, Anne Conway. Through the years of pain that her headaches caused her, Lady Conway came to see the soul and the world as united rather than divided, with spirits shuttling between them as angelic messengers. It was not a popular philosophy in the 1600s, an era better suited to Willis's clocks and alembics. Centuries have passed since Thomas Willis and Lady Conway last saw each other, when the doctor sat by the noblewoman's bed in a darkened room and discussed animal spirits and the nature of pain. But long after death, they are due for a reunion, to work together to make sense of the soul.

Dramatis Personae

Thomas Aquinas (1224 or 1225–1274) Italian theologian and philosopher. Helped to incorporate Aristotle into medieval Christian thought and establish natural philosophy.

Aristotle (384–322 B.C.) Greek philosopher. His system of knowledge dominated Europe until the seventeenth century. Believed the brain serves mainly to cool the heart.

Ralph Bathurst (1620–1704) English physician. Oxford virtuoso and early follower of William Harvey. Close friend and medical partner of Thomas Willis.

Robert Boyle (1627–1691) Irish-born natural philosopher. Helped transform alchemy into chemistry and establish the experimental tradition of modern science.

Charles I (1600–1649) King of Great Britain and Ireland (1625–1649). Son of James I and patron of William Harvey. Fought Parliament in the English Civil War. Executed.

Charles II (1630–1685) King of Great Britain and Ireland (1660–1685). Son of Charles I. Forced into exile in the English Civil War and restored to the throne after the fall of the Protectorate.

Anne, Viscountess Conway (1631–1679) Author of a copious correspondence and of the posthumously published *The Principles of the Most Ancient and Modern Philosophy* (1690).

Nicolaus Copernicus (1473–1543) Polish astronomer. Proposed that the Earth is a planet and that all the planets revolve around the sun.

Oliver Cromwell (1599–1658) Leader of parliamentary forces in the English Civil War and later Lord Protector of England.

Richard Cromwell (1626–1712) Son of Oliver Cromwell. Served as Lord Protector of England 1658–1659 before being driven from power.

René Descartes (1596–1650) French mathematician and philosopher. Father of modern philosophy.

Empedocles (ca. 490–430 B.C.) Greek philosopher. Best known for his cosmology based on four elements.

Epicurus (341–270 B.C.) Greek philosopher. Argued that the world is composed of atoms. His philosophy was reviled in the Middle Ages but revived by Pierre Gassendi.

Hieronymus Fabricius (1537–1619) Italian surgeon and anatomist. Taught William Harvey at the University of Padua.

Thomas Fairfax (1612–1671) Commander of parliamentary army during the English Civil War, led siege of Oxford.

John Fell (1625–1686) Fellow soldier and friend of Thomas Willis during English Civil War. Brother of Willis's first wife. Went on to become bishop of Oxford after the Restoration.

Galen (A.D. 129–ca. 199) Greek physician. His philosophy dominated European medicine until the seventeenth century.

Galileo Galilei (1564–1642) Italian natural philosopher. Helped found modern physics and astronomy.

Pierre Gassendi (1592–1655) French philosopher. Revived Epicurus's theory of atoms. Profoundly influenced Willis on the question of the soul.

William Harvey (1578–1657) English physician. Discovered the circulation of the blood and established physiology as an experimental science.

Joan Baptista van Helmont (1579–1644) Flemish physician and chemist. Pioneer of biochemistry who promoted mystical concepts about bodily ferments.

Franciscus Mercurius van Helmont (1614–1699) Son of Joan Baptista van Helmont. Physician and mystic. Treated Anne Conway.

Thomas Hobbes (1588–1679) English philosopher. Founded political science and championed a materialist view of the mind.

Robert Hooke (1635–1703) English physicist. Joined the Oxford circle as a student and was employed by Robert Boyle. Later became curator of experiments for the Royal Society. Among his many accomplishments, he published *Micrographia* in 1665.

James I (1566–1625) King of Scotland as James VI (1567–1625) and of England (1603–1625). Father of Charles I.

James II (1633–1701) Succeeded his brother Charles II as king of Great Britain and Ireland (1685–1688).

Edmund King (1629–1709) English physician. Royal surgeon and assistant to Thomas Willis in his London research.

William Laud (1573–1645) Archbishop of Canterbury (1633–1645). Political and religious advisor to King Charles I. Established conservative education at Oxford University. Executed.

John Locke (1632–1704) English philosopher. Helped usher in the Enlightenment with his account of the nature of human reason. Studied medicine under Thomas Willis.

Richard Lower (1631–1691) English physician and physiologist. Junior partner of Thomas Willis and later a prominent London physician. Performed dissections of brains with Willis in Oxford and also experimented with blood transfusions.

Lucretius (ca. 96–ca. 55 B.C.) Roman poet and philosopher. Known for his poem *De rerum natura* (*On the Nature of Things*), the fullest account of Epicurus's philosophy in classical writing.

Marin Mersenne (1588–1648) French natural philosopher and mathematician. Friend and correspondent of Descartes.

Henry More (1614–1687) English antimaterialist philosopher. Revived Plato's philosophy and helped introduce Descartes to England. Lifelong friend of Anne Conway.

Richard Overton (fl. 1646) English pamphleteer. Leveller. Championed mortalism in *Man Wholly Mortal*.

John Owen (1616–1683) English Puritan minister. Aide to Oliver Cromwell and vice-chancellor of Oxford University (1652–1659).

Paracelsus (1493–1541) Swiss physician. Incorporated alchemy into Renaissance medicine and championed a mystical view of life.

William Petty (1623–1687) English physician and political economist. Member of the Oxford circle, surveyor of Ireland, pioneer statistician.

Plato (ca. 428–348 or 347 B.C.) Greek philosopher. Established three-soul conception of the human body.

Anthony Ashley Cooper, first earl of Shaftesbury (1621–1683) English politician, led opposition in Parliament to Charles II. Patron of John Locke and patient of Thomas Willis.

Gilbert Sheldon (1598–1677) Archbishop of Canterbury (1663–1667). Thomas Willis's patron after the Restoration.

George Starkey (d. 1665) American-educated alchemist. Helped train Robert Boyle. Died of the plague.

Thomas Sydenham (1624–1689) English physician. Championed observation over obsolete theories. Strongly influenced John Locke.

Andreas Vesalius (1514–1564) Flemish anatomist. Offered the first serious challenge to Galen's scheme of human anatomy.

John Wallis (1616–1703) English mathematician and member of the Oxford circle. Broke royal codes for Parliament during the English Civil War.

Seth Ward (1617–1689) English astronomer. Member of the Oxford circle. Bishop of Salisbury.

John Wilkins (1614–1672) English mathematician. Played a leading role in establishing the Oxford circle and the Royal Society.

Thomas Willis (1621–1675) English anatomist and physician. Founder of neurology.

Christopher Wren (1632–1723) Best known as England's greatest architect. Joined the Oxford circle as a student and drew the illustrations that accompany Thomas Willis's *The Anatomy of the Brain and Nerves.*

Notes

A General Note on Sources

The people whom I describe in this book do not yet add up. Historians of science continue to produce fresh and important studies of Boyle, Descartes, Harvey, Locke, Willis, and all the other natural philosophers who gave us the modern brain. They uncover lost letters and manuscripts, they scrape away the bias of slanted histories, and they bring figures from the wings to center stage. But all this information does not yet cohere into a consistent shape. Arguments flare over hidden influences and agendas, about codes and loaded phrases. It can be tempting to ignore these debates and simply read the original writings of the natural philosophers, in search of direct revelations. But that doesn't bring a clear signal either. In some cases, there isn't enough to read—Thomas Willis wrote two hundred letters to his friend Ralph Bathurst that have been lost, for example. As for someone like Robert Boyle, who wrote three million words over his lifetime, there is too much. A dozen different Boyles can be crafted from his writings, each backed up by hundreds of pieces of evidence.

Because I set out to write a layman's account of some of the brain's history, I did not spend a lot of time in this book detailing the evidence and interpretations that were available to me. Instead, I tried to give accounts of these people that were consistent with the current consensus among historians. I do not agree with all interpretations, and I think that some debates have been unnecessarily polarized by false dichotomies. In these cases, I have chosen what seems to me to be the best arguments about an individual.

Historians have slowly been coming to appreciate the significance of Thomas Willis over the past century or so. Max Neuburger's 1897 history of neuroscience (reprinted in English in 1981) set him firmly at the start of the experimental tradition that flowers so luxuriantly today. Feindel (1962) argued that Willis was actually the founder of neurology, and his splendid 1965 edition of *Cerebri anatome* included not only spectacular reproductions of Wren's engravings but a long historical introduction. The first full-length biography was written by Hansreudi Isler (1968), and while it lacks the sophistication of more recent works, it is still loaded with important insights. Kenneth Dewhurst has edited Willis's lectures and casebooks, and his introductory essays and notes are indispensable additions to Willisiana. More recently, a new generation of historians has taken a fresh look at Willis as part of a network of virtuosi who led the scientific revolution. Robert Frank's magisterial *Harvey and the Oxford Physiologists* (1980) has set the standard for these studies. Robert Martensen has followed up on Frank's work with important studies of how Willis's social position influenced both his science and its reception. His upcoming book, to be published by Oxford University Press, will doubtless bring together the research that has previously appeared in his dissertation and subsequent journal articles.

The spelling in quotations has generally been modernized.

Introduction: A Bowl of Curds

Page

3 *the physik garden:* A list of plants found in the Oxford botanical garden can be found in Stephens 1658.

4 *Knives, saws, and gimlets:* The tools described here are based on the frontisplate of the Amsterdam 1667 edition of Willis's *Cerebri anatome,* as well as the list of standard dissecting equipment from a popular textbook of the time, Crooke 1631. The social gathering in which Willis did his work is analyzed in Martensen 1999.

5 *"like a dark red pigge":* Anthony Wood 1891.

5 *"this lax pith":* quoted in Henry 1989, p. 101.

5 *"No spirit, no God":* More 1653.

6 *were alchemists:* I refer to Thomas Willis in this book as an alchemist. Since he never claimed to turn base metals into gold, this may seem unjustified. But alchemy was much more than transmutation. It included a medical tradition, which came to be known as iatrochemistry, of which Willis was one of the leading figures in the seventeenth century. William Newman, in his invaluable *Gehennical Fire,* discusses the tricky question of what to call the work of people such as Willis. Is it alchemy, chemistry, or chymistry? He concludes, "When the term iatrochemistry appears in this book, therefore, let no one assume that it refers simply to the disinterested production of pharmaceuticals. As a chemical physician one could be—and usually was—an alchemist in every sense." I follow Newman on this matter. Willis did not use the spiritual, millennial rhetoric of Puritan alchemists, but as Hawkins points out, he did hope that his medicine would help heal the spirit of England. And his secrecy with his recipes certainly is in keeping with the alchemists' traditions.

Chapter One: Hearts and Minds, Livers and Stomachs

Page

9 *The oldest records of the procedure:* Egyptian techniques are described in Finger 1994; Persaud 1984; Boyle 2002.

10 *it resides around the heart:* Empedocles 2001.

10 *"all the senses":* Longring 1993, p. 58.

10 *Greeks in general were reluctant:* Edelstein 1935.

11 *"of another nature":* Longrigg 1993, p. 128.

11 *"the part of the soul":* Cunningham 1997, p. 12.

11 *"is endowed with courage":* Cunningham 1997, p. 11.
12 *the world's first biologist:* Longrigg 1993.
12 *stillborn babies:* Shaw 1972.
12 *a ladder from low to high:* van der Eijk 2000.
13 *a far more logical place:* Clarke 1963; Clarke and Stannard 1963.
13 *Herophilus and Erasistratus:* Von Staden 1989; Longrigg 1988.
14 *Galen:* Galen and May 1968; Rocca 1998; Temkin 1973; Tieleman 1996.
16 *residing in the blood:* Deut. 12:23; Lev. 17:11.
16 *the empty ventricles of the head:* Matthews 2000.
16 *speech came out of the chest:* Tieleman 1996.
17 *an open heart:* Stevens 1997.
17 *the soul was no different:* Longrigg 1993; Osler 1994, p. 63; Claus 1981.
17 *"Death is therefore nothing":* Pullman 1998, p. 43.
18 *Dante spoke for many:* see *Inferno*, Book X, line 14.
18 *Thomas Aquinas:* Michael and Michael 1989a.
19 *"the might of God to be marveled":* Cunningham 1997, p. 52.
19 *the immortal soul's tools:* Bono 1995.
19 *a series of giant charts:* French 1999.
20 *Vesalius found two hundred pieces:* Nutton 2002.
20 *"Show them to me":* Cunningham 1997, p. 111.
21 *"Lest I come into collision":* O'Malley 1997.
21 *the temperament:* Jackson 1986.
21 *Too much black bile:* Jobe 1976.
22 *the casebooks of Richard Napier:* MacDonald 1981; MacDonald 1990.
22 *"much afflicted with mopish melancholy":* MacDonald 1981, p. 31.
22 *"'Kiss my arse'":* MacDonald 1981, p. 202.
23 *Old friends:* Walker 2000.
23 *wound up on its list:* Shea 1986.
23 *conservative theologians:* Walker 1985.
23 *Aristotle could now inspire dangerous ideas:* Mercer 1993.
23 *Pietro Pomponazzi:* Michael and Michael 1989b.
23 *"will always be prepared to die":* Randall 1962, p. 79.
23 *Rome condemned Pomponazzi:* Kessler 1988; Osler 1994; Michael 2000; Michael and Michael 1989b.

Chapter Two: World Without Soul

Page
25 *Nicolaus Copernicus:* Hall 1983; Dear 2001; Lindberg 1992.
25 *Aristotle's picture of the cosmos:* Lindberg 1992.

27 *Galileo Galilei:* Drake, Swerdlow, and Levere 1999.
28 *"The Anatomy of the World":* lines 203 and 211.
28 *It had to part ways with Aristotle:* Hine 1984.
28 *Marin Mersenne:* Dear 1988; Dear 1991.
28 *take the soul out of nature:* Gaukroger 1995.
28 *A plucked lute string:* Vickers 1984.
29 *Gassendi claimed that salt could dissolve:* Clericuzio 2000.
29 *"all this to the degree":* Sarasohn 1985, p. 366.
29 *a soul made of atoms:* Michael 2000.
29 *it could think:* Michael and Michael 1989a.
30 *René Descartes:* Gaukroger 1995; Vrooman 1970.
30 *"All the doctors":* Gaukroger 1995, p. 16.
33 *The astronomer Johannes Kepler:* Crombie 1967.
34 *the butcher shops:* Wilson 2000.
34 *"We shall have no more occasion":* Descartes 1972, p. 115.
36 *"There is only one soul in us":* Wright 1980, p. 238.
37 *The pineal gland:* Smith 1998.
37 *"The mind," he declared, "can operate":* Conway 1996, p. xvi.
38 *"I wouldn't want to publish a discourse":* Gaukroger 1995, p. 290.
38 *"He has not permitted any falsity":* Harrison 2002b.
38 *Here was a divine guarantee:* Dear 2001.
39 *Riots broke out:* French 1999, p. 264.
39 *Dutch theologians:* Wilson 2000.
40 *Jesus' soul joined with the bread:* Dear 1991.

Chapter Three: Make Motion Cease

Page
43 *struck by their intelligence:* Examples of animal intelligence are found in Willis 1683.
45 *"bower of plenty":* Schama 2001, p. 22.
45 *"Kings sit in the throne":* Schama 2001, p. 27.
46 *it was made a crime for anyone other than the king:* Bloch 1973.
46 *"This royal pair":* Carew 1949.
47 *"threatened to rival the siege of Troy":* Cooke 1975, p. 184.
47 *"the greatest place":* Tinniswood 2001, p. 7.
48 *Laud did believe in learning:* Tyacke 1978.
48 *"a knowing woman":* Wood 1891.
49 *A kitchen was like a clinic:* Wear 2000, p. 54.
50 *"cast a corpse into the salty sea":* Principe 2000, p. 61.

51 *Paracelsus:* Debus 1976; Pachter 1951; Pagel 1972; Trevor-Roper 1985; Webster 1982.
51 *"They are vainglorious babblers":* Pachter 1951.
53 *Yet even Luther was too conservative:* Trevor-Roper 1998.
53 *a fraction of his writings:* Webster 2002.
53 *The Catholic Church banned his books:* Trevor-Roper 1998.
53 *"Prince of Homicide Physicians":* Trevor-Roper 1998, p. 230.
54 *the spirit of the world itself:* Clericuzio 1994.
54 *He organized the* London Pharmacopoeia: Trevor-Roper 1998.
54 *the hero of the outcast:* Webster 1975.
54 *"If I find anything":* MacDonald 1981, p. 188.
55 *"his mistress would oftentimes":* Aubrey 1898, p. 303.

Chapter Four: The Broken Heart of the Republic

Page
58 *"By God, for not an hour":* Ashley 1990.
58 *"The turning of a straw":* Hill 1972, p. 19.
59 *Tensions flared:* Details of Oxford on the eve of war can be found in Roy 1992; Roy and Reinhart 1997.
59 *a "tolerable livelihood":* Frank 1980, p. 106.
59 *this obsolete education:* For an example of standard anatomy instruction in pre-war England, see Crooke 1631.
60 *"This disease grew so grievous":* Willis 1681b.
61 *first clinical description of typhus:* Frank 1990.
61 *Only if order was restored:* Hawkins 1995.
61 *a royalist pamphlet:* Bloch 1973, p. 209.
62 *"dark, nasty room":* Wood 1891.
63 *buffoons and atheists:* Thomas 1969, p. 94.
63 *a fire broke out:* Porter 1984.
63 *"Dead hogs, dogs, cats":* Capp 1994.
64 *"As for the young men of the city":* Wood 1891.
65 *"He did delight to be in the darke":* Aubrey 1898.
66 *he followed Aristotle as a leader:* Cunningham 1997, p. 184.
68 *"The heart is not only the origin":* Clarke and O'Malley 1996, p. 25.
69 *one of the few physicians in London who could cure devils in the head:* Hunter and Macalpine 1957, p. 137.
69 *"the first principle":* Harvey 1981, p. 296.
69 *"It is clear that all sensation":* Harvey 1981, p. 296.
69 *"That sensation as well as movement":* Harvey 1981, p. 248.

69 *the world-spirit infused the body with spirits of its own:* Walker 1985.
69 *"I have never been able to find any such a spirit anywhere":* Harvey 1981, p. 375.
69 *Blood, Harvey believed, did everything:* Bono 1995.
69 *"serve as a common subterfuge":* Bono 1995, p. 86.
70 *"the heart of animals is the foundation":* Harvey 1993, p. 1.
71 *"But by Hercules!":* Harvey 1993, p. 20.
72 *"My whole life, perchance":* Davis 1973, p. 142.
72 *pregnant does:* Keynes 1966.
72 *"himself much delighted":* Harvey 1981, p. 336.
72 *"It was then so small":* Harvey 1981, p. 359.
72 *"I was greatly afraid":* Harvey 1993, p. 9.
73 *They opened his book:* French 1994.
73 *"fell mightily in his practice":* Aubrey 1898.
74 *"I immediately saw a vast hole":* Harvey 1981, p. 250.
74 *"I was almost tempted":* Harvey 1993, p. 23.
75 *"I wish I could perceive the thoughts":* Harvey 1981, p. xxiii.
75 *"He pursued most diligently":* Payne 1957, p. 163.
75 *"While I did attend upon his most Serene Majesty":* Harvey 1981.
75 *"a great mischievous baboon":* Keynes 1966, p. 433.
76 *"he observed more people die of grief":* Hunter and Macalpine 1957.
77 *Oliver Cromwell:* Two useful biographies of Cromwell are Firth 1953; Hill 1970.
77 *"I had rather that Mahometanism":* Firth 1953, p. 300.
77 *"every man by nature":* Richard Overton, quoted in Schaffer 1983, p. 118.
78 *"to turn the world upside down":* Webster 1975, p. 180.
78 Man Wholly Mortal: Overton 1655.
79 *Mortalists:* Burns 1972.
79 *"a pig from Epicurus's stye":* Jones 1989, p. 198.
79 *Guy Holland:* Holland 1653.
80 *"the nurseries of wickedness" and "cages of unclean birds":* Webster 1973.
81 *"Now the Sun is unarmed":* From "A mock-song," by Richard Lovelace.

Chapter Five: Pisse-Prophets Among the Puritans

Page
84 *The king had reached the Scottish forces:* Ashley 1990.
84 *Charles and his captors were mobbed:* Bloch 1973.
84 *Stroker:* Thomas 1971, p. 197.

86 *"a plain Man:"* Meyer and Hierons 1965b, p. 146.
86 *"a sword in a blind man's hand":* Willis 1981, p. 10.
86 *"the filthiness and soot":* Quoted in Isler 1968.
86 *"to say a house consists of wood":* Debus 1976, p. 523.
87 *Joan Baptista van Helmont:* For biographical information, see Pagel 1970; Pagel 1982.
87 *"It dawned on me":* Debus 1970, p. 17.
88 *hydrochloric acid:* Pagel 1956.
89 *"the devil's deceit":* Pagel 1970, p. 254.
89 *The salve could waken the blood's soul:* Meier 1979.
90 *"ordained in these last times":* George Thomson quoted in Webster 1971, p. 156.
90 *"Pause traveler!":* French 1999, p. 249.
91 *thunder and lightning:* Guerlac 1954.
91 *an inner violence:* Paster 1997.
91 *"a vehement crash":* Willis 1681a, p. 2.
91 *this sort of flameless explosion:* Frank 1974.
92 *Thomas Willis remained:* Hawkins 1995.
92 *"An almost general riddance was made":* Allestree and Fell 1684, p. 2.
92 *"harpies and stinking fellows":* Shapiro 1969, p. 84.
92 *John Wilkins:* For a biography of Wilkins, see Shapiro 1969.
92 *"a very mechanical head":* Aubrey 1898.
92 *a flying-chariot:* Wilkins 1638, p. 128.
93 The New Planet: Ross 1646.
93 *"if there be nothing able":* Wilkins 1638.
93 *"It is an excellent rule":* Wilkins 1638.
93 *virtuosi:* Houghton 1942.
94 *Wren:* Jardine 2002.
94 *"Either God or Copernicus":* Bennett 1982.
95 *"a most addicted client":* Bennett 1982, p. 17.
95 *new microscopes:* Bennett 1982, p. 73.
95 *a compass needle:* Bennett 1982, p. 47.
95 *intricate beehives out of glass:* Hartlib 1655.
95 *"our chymist":* Frank 1974.
95 *William Petty:* No modern full-length biography of William Petty exists. But see Strauss 1954; Petty 1927; Aspromourgos 1996; Alexander 2000b; Adams 1999.
96 *Thomas Hobbes:* See Gert 1996; Martinich 1999; Mintz 1962; Peters 1956; Rogers 2000; Shapin and Schaffer 1985; Skinner 1966; Skinner 1969; Tuck 1992.

97 *"For what is the heart":* Peters 1956, p. 22.
97 *"It is manifest":* Hobbes 1994, p. 214.
97 *"brain or spirits":* Hobbes 1994, p. 23.
97 *The brain pushed back:* Shapiro 1973.
97 *"the fountain of all senses":* Hobbes 1994, p. 216.
97 *So too were memories:* Sutton 1998, p. 150.
97 *Stirring up the spleen:* Hobbes 1994, pp. 28 and 222.
98 *Reason was not the work of an immaterial soul:* James 1997.
98 *"no less than that whereby a stone":* Sarasohn 1985, p. 369.
98 *"I put for a general inclination":* Leviathan, pt. 1 chapter 11.
99 *"For unless I am very much mistaken":* Martinich 1999.
99 *"who always examines everything":* Skinner 1966, p. 160.
100 *letters from the commander of the local garrison:* Tyacke 1997, p. 544.
101 *"sought truth":* Strauss 1954, p. 40.
101 *the bladder of an ox:* Willis 1981, p. 34.
101 *"The most mysterious and complicated enginry":* Petty 1927, vol. 2, p. 172.
101 *"a most excellent mathematician":* Webster 1969, p. 367.
101 *"I cannot believe his principles":* Webster 1969, p. 368.
102 *A few hundred royalists:* Ashley 1990.
102 *"we will cut off his head":* Hill 1970, p. 103.
104 *"was night and day":* Willis 1981, p. 141.
105 *"Whilst talking":* Quoted in Eadie 2003b.
105 *"I have known some":* Willis 1681a.
108 *"She was fat and fleshy":* Petty 1927, p. 161.
108 *"And while she was hanging":* Petty 1927, p. 158.
109 *"with the coffin wherein she lay":* Watkins 1651.
109 *"my endeavours":* Frank 1997, p. 546.
110 *"as if there had been":* Petty 1927, p. 157.
110 *"her memory was like a clock":* Watkins 1651, p. 6.
111 *"there was, in the bottom of the vessel":* Willis 1981, p. 118.
111 *"as in mechanical things":* Willis 1681c, p. 39.
112 *"to enlarge my trade":* Strauss 1954, p. 110.
112 *"the Body Natural":* Strauss 1954, p. 192.
114 *lived just a few doors away:* Tyacke 1997, p. 754.
114 *"discharge eight cannons":* Wood 1891.
114 *"a peace more cruel than any war":* Willis 1684.
114 *Richard Allestree:* Allestree and Fell 1684.

Chapter Six: The Circle of Willis

Page

117 *Parliament classified Harvey as a delinquent:* Harvey 1981, p. xxii.

117 *"a pretty young wench":* Aubrey 1898.

118 *"the mere office of a Midwife":* Harvey 1653.

119 *Only a soul:* For Harvey's view on mechanism and development, see Brown 1968.

119 *"the soul and its affections":* Hunter and Macalpine 1957, p. 136.

119 *"I should do nothing":* Payne 1957, p. 163.

119 *"till he doted":* Quoted in Frank 1974, p. 171.

119 *financed a magnificent library:* Frank 1979, p. 11.

121 *the earliest known clinical description of influenza:* Bates 1965.

121 *"Sitting oftentimes by the sick":* Willis 1681b.

122 *Hooke:* Andrade 1950; Jardine 2003.

122 *flying machines:* Andrade 1950.

123 *"I pity to see so many young heads":* Jones 1961, p. 121.

123 *"mere Moral men":* Webster 1975, p. 172.

123 *"were good at two things":* Frank 1980, p. 56.

123 *"daily converse with the Heathens":* Webster 1975, p. 187.

123 *"what is it else but a confused Chaos":* Webster in Debus 1970, p. 15.

124 *"ignorant pagan":* Debus 1970, p. 73.

124 *"I dare truly and boldly say":* Debus 1970, p. 71.

124 *"I say you are no Parliament":* Firth 1953.

125 *"You are at the edge of the promises and prophesies":* Firth 1953, p. 324.

125 *"When the microscope":* Ward in Debus 1970, p. 35.

125 *"Chymicall society":* Debus 1970, p. 36.

125 *"Surgery as well as Physick":* Debus 1970, p. 35.

126 *"the Romans will come":* Debus 1970, p. 65.

127 *Oxford's scholars responded:* For the persecution of Oxford Quakers, see Braithwaite 1955.

127 *"are nothing but the distempers of a disaffected brain":* Quoted in Johns 1996, p. 143.

127 *"Men therefore must now be sober to God":* Hill 1972, p. 226.

128 *"He walked much and contemplated":* Aubrey 1898.

128 *"power of this kind":* Tuck 1992.

128 *Hobbes published* Leviathan: For the reception of the book, see Laver 1978, Bloch 1973.

129 *"the kingdom of fairies":* Tuck 1992, p. 111.

129 *a palsy:* Hobbes's medical condition is considered in Kassler 2001.

129 *"I returned to my homeland":* Quoted in Martinich 1999.

129 *"if anyone objected":* Quoted in Martinich 1999, p. 219.
129 *"vomited up the condemned opinions":* Quoted in Martinich 1999, p. 258.
130 *"numb with horror":* Ward and Wallis quoted in Martinich 1999, p. 279.
130 *certainly did not want to be confused with the likes of Hobbes:* Henry 1986.
131 *it would have stalled:* For the impact of thinkers such as Descartes and Hobbes, see Finger 2000.
131 *England's aristocrats had been amusing themselves:* Eamon 1994.
131 *Robert Boyle:* Hunter 2000.
131 *Richard Boyle:* Canny 1982.
132 *in Padua he saw arteries:* Oster 1989.
133 *"They must henceforward maintain themselves":* Macintosh 2002.
133 *"the foulest churl":* Quoted in More 1944, p. 54.
135 *George Starkey:* The definitive biography of this remarkable man is Newman 1994.
136 *Starkey needed to attract the attention of powerful, wealthy patrons:* For the social function of Eirenaeus Philalethes, see Newman 1994; Martensen 2002.
137 *Starkey inherited the tradition:* For Starkey's link to atomist alchemy, see Newman 2001.
137 *"Let that beastly Epicure's mouth":* Crooke 1631, p. 8.
137 *the invisible scent of a partridge:* Macintosh 1991.
137 *"what a multitude":* Oster 1989, p. 154.
138 *"much disorder the affairs of Mankind":* Hunter 1990a, 407.
138 *"conversing with dead and stinking Carkases":* Kaplan 1993, p. 38.
138 *"squander away a whole afternoon":* Hunter 1999, p. 261.
138 *"is ever an impediment":* Alexander 2000a, p. 109.
139 *"a very ingenious experimenting gentleman":* Frank 1980.
139 *"be a means to quicken and direct our enquiries":* Maddison 1969, p. 81.
139 *millions of his words:* Kahr 1999.
139 *an alchemist's code:* Principe 1992.
139 *lost five hundred experiments:* Macintosh 2002.
139 *"speaks French and Italian very well":* Quoted in Hunter 2000, p. 59.
140 *made his point with saltpeter:* Boyle 1999, vol. 2, p. 93ff.
140 *"limpid Liquor":* Boyle 1999, vol. 2, p. 94.
142 *"I dare speak":* Quoted in Macintosh 1991, p. 200.
142 *"It more sets off the wisdom of God":* Henry 1986, p. 354.
143 *pure reason alone:* Harrison 2002a, 2002b.

143 *the grammar of God's language:* McGuire 1972.
144 *a visit to Harvey:* Bylebyl 1982.
144 *"I have not been so nice":* Boyle 1999, vol. 3, p. 211.
144 *"I look not on a human body":* Boyle 1996, p. 127. For Boyle's views on bodies and machines, see Giglioni 1995; Clericuzio 1994.
144 *Life and death:* Kaplan 1993.
145 *"knows both that air":* Quoted in Anstey 2001, p. 490.
145 *"This hypothesis":* Boyle 1996, p. 143.
145 *"the bare remembrances of a loathsome Potion":* Boyle 1999, vol. 4, p. 442.
145 *"a considerable quantity of Briny Liquor":* Boyle 1999, vol. 4, p. 443.
145 *The human body was like a musket:* Boyle 1999, vol. 4, p. 445.
145 *"It must be a strange aggregate":* Quoted in Anstey 2001, p. 510.
145 *"will, I confess, not only instruct me":* Quoted in Anstey 2001, p. 497.

Chapter Seven: Spirits of Blood, Spirits of Air

Page
147 *"I thought best, the common acception":* Willis 1681b, p. 1.
147 *"All Natural effects":* Quoted in Debus 1976, p. 523.
148 *earth, water, salt:* Willis may have drawn these principles from the English physician Francis Glisson, as noted in Clericuzio 2000.
148 *"Aetherial Particles":* Quoted in Clericuzio 1994, p. 60.
148 *a ferment:* For a general discussion of Willis's ferment, see Hall 1969.
148 *"without recourse to occult qualities":* Willis 1681b, p. 2.
149 *Van Helmont had claimed:* Brown 1968.
149 *"We are not only born":* Quoted from Willis 1681b.
150 *"This, I fear, may prove":* Quoted in Clericuzio 2000.
150 *"a man uncommonly skilled":* Oldenburg 1965.
151 *"the lechery of the wanton mind":* Edmund O'Meara quoted in Martensen 1993, p. 185.
151 *"We should rather reject":* Martensen 1993.
152 *"The opening of a vein":* Quoted in Brown 1968, p. 158.
152 *a backward-looking revolutionary:* For a general discussion of Willis's medicine, see Brown 1970.
153 *the most active principle of nature:* Meier 1982.
153 *A glassy alembic:* The physician Sylvius also used the alembic metaphor in his writing on the brain around the same time as Willis. See Forrester 2002.

154 *the way anatomists normally examined:* For a typical case of horizontal sectioning in brain anatomy, see Vesling and Blasius 1666.
154 *"wander without a leader:"* Willis 1681b.
154 *Anabaptists were stirring up Oxford:* Jones 1961, p. 114.
155 *"Before I came to him":* Quoted in Firth 1953, p. 434.
155 *Religious sects swarmed:* Hutton 1985, p. 26.
155 *"Oh priests":* Simpson 1659, p. 1.
155 *"is only to turn out the landlords":* Conway 1930, p. 161.
156 *"This is a call for every uncircumcised Philistine":* Fox 1659.
157 *as far away as Cambridgeshire:* Frank 1980, p. 180.
157 *They poured blood from a vein:* Willis 1681b, p. 62.
158 *"We live at the bottom of an ocean of the element air":* Shapin 1996, p. 39.
158 *the moon caused ocean tides:* Wren and Boyle's test is described in Bennett 1982.
159 *a list of experiments:* Boyle 1999, vol. 1, p. 299.
160 *springy particles like an invisible wool:* Clericuzio 2000.
160 *a lark in the globe:* Hall 1969.
161 *a living fire:* Clericuzio 1993.
161 *"This invited us":* Boyle 1999, vol. 1, p. 293.
161 *"such as is offr'd to a Wind-Mill":* Boyle 1999, vol. 1, p. 294.
161 *"There is some use of the Air":* Boyle 1999, vol. 1, p. 295.
161 *particles mixed among the air:* Boyle 1999, vol. 1, p. 282.
162 *"little Veins & Arteries in them":* Bennett 1982, p. 78.
162 *an ulcerated intestine:* This sketch is reprinted in Doby 1973.
163 *"It seems that it is":* Willis 1681b.
164 *"liked his entertainment":* Schaffer 1998, p. 95.
165 *"When his vein was closed":* Willis and Pordage 1683.
165 *"an inferior Domestick of his":* Schaffer 1998, p. 96.
166 *"spirit of wine":* Gunther and Gunther 1920, vol. 13, p. 104.
166 *"there may be some way":* Gunther and Gunther 1920, vol. 13, p. 107.
167 *the city's fountains spouted claret and beer:* Fasnacht 1954.

Chapter Eight: A Curious Quilted Ball

Page:
170 Paradise Lost *and* Pilgrim's Progress: Hutton 1985, p. 156.
170 *atheism:* The best studies of atheism in Restoration England are Hunter 1985; Hunter 1990b.
171 *"no fools are so great":* Stillingfleet 1667, p. 7.

171 *"He had a very ill opinion":* Burnet's observations are quoted from Burnet 1823.
172 *a bastard at a royal ball:* Hutton 1989, p. 186.
172 *the King's Evil:* Schaffer 1998, p. 87.
173 *"has driven out all other philosophies and rules supreme":* Quoted in Isler 1968, p. 30.
173 auctorium: Larner 1987.
174 *"comment on the Offices":* Willis 1965.
175 *"At length we have had the opportunity":* Lower to Boyle, January 18, 1661, in Boyle 2001.
175 *Willis would then hold it:* For the impact of Willis's method, see Martensen 1995.
176 *the structure of the nerves:* Martensen 1995.
177 *"very fast and close with silk":* Lower to Boyle, Boyle 2001.
177 *the Circle of Willis:* For the history of the *rete mirabile* and the Circle of Willis, see Forrester 2002.
177 *a ferment distilled the spirits:* Clericuzio 1994.
178 *"we can scarce believe":* Willis 1965, p. 106.
178 *corpus striatum:* This area is now known commonly as the basal ganglia, a region of the brain crucial to organizing body movements. Parent 1986.
179 *"The Spirits inhabiting the Cerebel":* Willis 1965, p. 111.
179 *"cruel and horrid symptoms":* Willis 1965, 112.
180 *"Some think it is enough": Of muscular motion,* p. 35, in Willis 1681c.
180 *"explicate them according to the Rules": Of muscular motion,* p. 35, in Willis 1681c.
180 *carried commands:* Wallace 2003.
181 *an explosive fountain:* Canguilhem 1993.
181 *reflex:* Willis's contribution to the scientific understanding of reflex motion is discussed in Canguilhem 1993; Meyer and Hierons 1965b; Sherrington 1941.
181 *a maze far beyond anything found in any animal's brain:* Willis's comparative method is outlined in Bynum 1973.
181 *"These folds or rollings":* Willis and Pordage 1683, p. 76.
182 *a harmonious network of nerves:* Willis's establishment of the autonomic nervous system is detailed in Meier 1982.
183 Cerebri anatome: Willis 1664. For a brief overview of the background to *Cerebri anatome,* see O'Connor 2003.
184 *Charles was fond of the new science:* Mendelsohn 1992; Gunther and Gunther 1920, vol. 13, p. 95.

184 *"It is certain this New-Philosophy":* Jones 1989, p. 208.
184 *"Their first purpose":* Quoted in Strauss 1954.
185 *"When I speak":* Quoted in Jones 1989, p. 209.
185 *"truly acquainted":* MacDonald 1990, p. 67.
185 *evidence of witchcraft:* Clark 1997; Hunter 1990b, p. 395.
186 *"Nature does not only work":* Sawday 1995, p. 253.
186 *"We may perhaps be enabled":* Wilson 1988, p. 97.
186 *Bathurst presented the Society with the book:* Frank 1990.
186 *"the ornament of our nation":* Marchamont Nedham, quoted in Willis 1981, p. 166.
186 *"immortal work on the brain:"* Frank 1980, p. 198.
186 *anatomists across Europe:* A sign of Willis's universal authority was the 1666 edition of Vesling's anatomy textbook. In it, the author simply pasted charts from *Cerebri anatome*. See Vesling and Blasius 1666.
187 *twenty-three editions:* Diamond 1980.
187 *"the Harvey of the nervous system":* Brain 1963.
187 *"Once more your Sidley Professor":* Willis 1965.

Chapter Nine: Convulsions

Page
189 *"one of the learnedest":* Christie 1871, p. 294.
190 *The first clinical descriptions of migraine:* Isler 1986; Sacks 1992.
190 *a brilliant philosopher:* Popkin 1992.
190 *Anne Conway:* No book-length treatment of Conway exists. For more information on her life and philosophy, see Conway, 1930; Conway 1992; Conway 1996; Hutton 1996; Hutton 1997; Sherrer 1958; Skwire 1999.
190 *"She was of a most beautiful form":* Willis 1683, p. 122.
191 *More had built his life's work:* Hutton and Crocker 1990.
191 *"I should look upon Des-Cartes":* Webster 1969, p. 376. See also Henry 1989.
191 *experiments merely as demonstrations:* Webster 1969.
192 *"I brush the stars":* Conway 1930, p. 44.
193 *"You write like a man":* Conway 1930, p. 32.
193 *"to have a Physician":* Conway 1930, p. 30.
193 *"Yet he pretends":* Conway 1930, p. 73.
193 *"It hath pleased God":* Conway 1930, p. 181.
194 *a cure-all:* Boyle 1999, vol. 4, p. 392.
194 *"Having tried that kind of remedy":* Conway 1930, p. 91.

195 *Pain is the one unbroken thread:* Sherrer 1958.
196 *"I have some relief":* Conway 1992, p. 533.
196 *"Why does the spirit":* Hutton 1997, p. 229.
196 *"dull and stupid matter":* Hutton 1996, p. 239.
197 *angels, of course, being divine messengers:* Sutton 1998, p. 38.
197 *"For how can a dead thing":* Merchant 1980, p. 262.
197 *"If it be granted":* Hutton 1996, p. 241.
197 *"exclusion from the world":* Conway 1992, p. 533.
199 *His accounts of their headaches:* For Willis's work on headaches, see Isler 1986; Sacks 1992.
199 *To account for these pains:* Willis 1683, p. 108.
199 *a comet:* Wallis to Oldenburg, January 21, 1664/5; Oldenburg 1965, p. 353.
199 *calculate their paths:* Bennett 1982, p. 66.
200 *"Death, as it were":* Cowie 1970.
201 *"Therefore wine and confidence":* Willis 1691.
201 *"Go, you rascal":* Hutton 1985.
203 *a mysterious new kind of fever:* Clarke 1975.
203 *"to find out from her death":* Willis 1681a.
203 *new theories about epilepsy and other kinds of convulsions:* For an overview of this part of Willis's work, see Eadie 2003a, 2003b.
204 *"from unnatural states of the uterus":* Harvey and Whitteridge 1964.
204 *"wholly faultless":* Willis 1681a, p. 71.
204 *"imitating the type of hysterical attack":* Meyer and Hierons 1965a.
204 *Babylonians had considered it the work of a demon:* Longrigg 1993.
205 *epileptic seizure:* Temkin 1971.
205 *"He is not able to draw":* Willis 1681a, p. 12.
205 *"like a long train of gunpowder":* Tourney 1972.
206 *"This is precisely the modern view":* Brain 1963, p. 208; Temkin 1971.
206 *the psychological life of his patients:* Wright 1980.
206 *it tamed the racing animal spirits:* Veith 1965.
206 *"there is not a disease of the head":* Meyer and Hierons 1965b.
207 *"I believe there was never any such desolation":* Conway 1930, p. 277.

Chapter Ten: The Science of Brutes

Page
209 *"Men begin now everywhere":* Conway 1930, p. 278.
210 *The sheep survived:* Schaffer 1998. For other accounts of the blood

transfusion craze, see Schaffer 1998. See also Edmund King's letter to Boyle, November 25, 1667, in Boyle 2001.

211 *"was the subject":* Lower 1932.

211 *"improve his mental condition":* Lower 1932, p. 189.

212 *"If the blood moves":* Lower 1932, p. 62.

214 *stopped paying his dues:* Meyer and Hierons 1965b.

215 *"a mere muscle": Of muscular motion,* p. 39, in Willis 1681c.

216 *"He became so noted":* Hughes 1991, p. 86.

216 *the first truly great neuroscientist:* Feindel 1962.

217 *the brains belonged to England's ruling class:* This point is examined thoroughly in Martensen 1993; Martensen 1999.

217 *"the most admirable republics of bees":* Willis 1683.

217 *"endeavouring as much as they could":* Willis 1683, p. 3.

217 *"infinitely above the reach of our senses":* Wallace 2003, p. 87.

218 *"chords or strings":* Quoted in Cranefield 1961, p. 307.

218 *the brain of a monkey:* Bynum 1973.

219 *Monkeys' brains, he discovered, were even more like humans':* Bynum 1973.

219 *"came nearer the figure":* Willis and Pordage 1683.

220 *narcoleptics:* Lennox 1938.

220 *"watery deluge":* Willis and Pordage 1683, p. 134.

221 *"the reason of which is":* Willis 1965, p. 138.

221 *"we have noted little or no difference":* Willis 1683, p. 44.

221 *the rational soul:* Bynum 1973.

222 *"God, Angels, It Self, Infinity":* Willis 1683, p. 398.

222 *"having pitched its tents":* Willis 1683, p. 122.

223 *"We cannot here yield":* Willis 1683, p. 190.

223 *If the spleen stopped filtering the dregs:* Jackson 1986.

223 *carving new, contorted paths:* Jobe 1976.

224 *"Pleasant talk, or jesting":* Willis 1683, p. 194.

224 *"Furious Mad-men":* Willis 1683, p. 206.

224 *"most chiefly belong to the Rational Soul":* Willis 1683, p. 209.

224 *"stupidity" and "foolishness":* Cranefield 1961.

224 *Such categories:* See Goodey 1996; Goodey 1999 for fascinating challenges to a transhistorical concept of intellectual disability.

224 *"the moistest fire":* Hippocrates 1923, p. 281.

225 idiotae: Neugebauer 1978.

225 *"leaping forth":* Quoted in Cranefield 1961, p. 301.

225 *"those that are born of Plowmen":* Cranefield 1961, p. 296.

225 *To call a peasant an idiot:* Goodey, in press.

225 *Willis simply added a medical tone:* Goodey submitted.
226 *People could be born stupid or become stupid with age:* Cranefield 1961, p. 313.
226 *"In some Families":* Cranefield 1961, p. 311.
226 *special schools:* Martensen 1995.
227 *The proto-psychologists:* Lapointe 1970; Park and Kessler 1988; Hatfield 1995.
227 *"unconventional and unestablished":* Quoted in Clarke 1975, p. 291.
227 *"After the death of my dear wife":* Willis 1683, preface.
228 *the most complete account:* This is the assessment of, among others, Rousseau 1991; Richards 1992, p. 75.
228 *limited to it:* Rousseau 1991.
231 *"aetherial man":* Quoted in Johns 1996, p. 157.
231 *a material vapor:* Johns 1996, p. 152.
231 *"I have attributed too much to natural causes":* Quoted in Elmer 1986, p. 5.
231 *the last Paracelsist books:* Webster 2002.
232 *"Here comes the bear to be baited!":* Quoted in Mintz 1962, p. 14.
232 *"Most of the bad Principles":* Mintz 1962, p. 135.
232 *"If this be true":* Stillingfleet 1667.
232 *The Oxford circle also tangled with Hobbes:* Dear 2001; Skinner 1969; Shapin and Schaffer 1985.
233 *"How this Experiment will be reconcil'd":* Boyle 1999, vol. 4, p. 303.
233 *"thereby delivered his Majesty's secrets":* Martinich 1999.
234 *an account of the English Civil War:* See the introduction to Hobbes 1990.
234 *"the learned Author of this difficult argument":* Anonymous 1672, p. 4071.
234 *"if thou wilt study thy own frame of Body":* Ramesey 1672, p. 129.
235 *"They are mad":* Ramesey 1672, p. 41.
235 *"superstition and the despair of eternal salvation":* Willis 1683, p. 200.
235 *"wholly perverted from the use of right reason":* Willis 1683, p. 200.
235 *Priests and bishops:* For examples of religious uses of Willis's work, see Frank 1980; Frank 1990.
235 *a physical disorder:* For discussions of the medicalization of religion, see Porter 1983; Johns 1996; MacDonald 1981.

Chapter Eleven: The Neurologist Vanishes

Page
237 *"as secret as though I were its custodian":* Sachs to Oldenburg, Octo-

ber 29, 1671, p. 324 in Oldenburg 1965. For other requests, see Oldenburg to Sachs, December 22, 1671, p. 418 in Oldenburg 1965.

238 *"mechanical means of the workings of Medicines":* Quoted in Brown 1970.

238 *the book that the Civil War had stopped William Harvey from writing:* Martensen 2002.

238 *"physick being perfected":* Willis 1684.

238 *"as oil from a lamp":* Willis 1684, vol. 2, p. 27.

238 *diabetes:* For Willis's achievements in *Rational Therapeutics,* see Frank 1990.

238 *"altogether frivolous":* Willis 1684, volume 2, page 141.

239 *"Thou knewst the wondrous art":* Williams's poem appears in Willis 1684.

240 *When Locke came to Oxford:* Milton 1994, 2001.

241 *hundreds of pages:* Locke's notes were published in Dewhurst 1980.

242 *specimens which he pressed:* Frank 1973.

242 *"some small subtile parcels":* Dewhurst 1984.

244 *"The selfsame phenomena":* Stevenson 1965.

244 *intermittent fevers:* Wilson 1993.

244 *"than hath bin":* Dewhurst 1966, p. 39.

245 *"It is my nature to think":* Dewhurst 1966, p. 43.

246 *"Sydenham and some others in London":* Frank 1974.

246 *"All that Anatomie":* Dewhurst 1966.

246 *"The brain is the source":* Dewhurst 1966.

246 *"the imperfection of words":* Locke 1965 (Book 3, chapter 9, section 16).

247 *Thomas Willis paid a visit:* Anonymous 1683.

247 *Lord Ashley had used Willis's services in the past:* Dewhurst 1964.

247 *"We should do well":* Locke 1965 (Book 4, chapter 16, section 4).

248 *"had nothing to do with the good of men's souls":* Quoted in Cranston 1957, p. 112.

251 *"plain historical method":* Sanchez-Gonzalez 1990.

252 *To refuse to believe anything:* Locke 1965 (Book 1, chapter 1, section 5).

252 *"Hypotheses," Locke wrote, "if they are well made":* Locke 1965 (Book 4, chapter 12, section 13).

252 *"a little casque":* Cranston 1957, p. 310.

253 *"prescribing such intolerable pain":* Quoted in Clarke 1975, p. 302.

253 *"He would think of it":* Quoted in Clarke 1975, p. 301.

254 *"'Tis impossible to conceive":* Quoted in Hunter 1995, p. 118.

254 *"I am ready for the great experiment":* Quoted in Shapiro 1969.

255 *"He has been often heard to complain":* Quoted in Hunter 1995, p. 46.

256 *Willis's neurology runs through* An Essay: The relationship between Locke's philosophy and Willis's neurology is discussed in Isler 1968; Rousseau 1976; Wright 1991; Martensen 1993; MacDonald 1990; Dewhurst 1984.
256 *blank sheet of paper:* Cranefield 1961; Cranefield 1970.
256 *"trains of motions":* Locke 1965 (Book 2, chapter 33, section 6).
256 *"in which, whilst they flow":* Willis 1683, p. 203.
257 *"a kind of imprisoned angel":* Macintosh 2002.
257 *"character of a Christian philosopher":* Shapin 1991.
257 *Boyle was so desperate:* Boyle's interviews with Stillingfleet and Burnet are detailed in Hunter 2000.

Chapter Twelve: The Soul's Microscope

Page
263 *"You are a doctor":* The questions quoted from Greene's study in this chapter are listed at http://www.sciencemag.org/cgi/content/full/293/5537/2105/DCI.
266 *the boar's semen:* Willis 1683, p. 30; Frank 1980, p. 183.
266 *"to carry the most pure flower":* Willis 1683, p. 30.
266 *"The Epitome":* Willis 1683, p. 22.
266 *wise fathers:* Goodey in press.
267 *within cells:* See Regev and Shapiro 2002; Holcombe and Paton 1998; Loewenstein 1999.
268 *Life evolves:* Schneider 2000; Adami 2000.
268 *a fail-safe antidote to atheism:* Bynum 1973.
268 *"like Leather burnt with the fire":* Wallis 1666, p. 222.
269 *the concept of animal spirits collapsed:* Pera 1992; Clower 1998.
269 *The nerves:* For an introduction to neurons and neuroscience, see Purves and Williams 2001.
270 *the ancient biochemistry of microbes:* Evidence of the evolutionary roots of neurons can be found in Chen et al. 1999; Anderson and Greenberg 2001; Anderson 1990.
270 Aplysia: See Weeber and Sweatt 2002; Albright et al. 2000.
271 *vertebrate evolution:* Butler and Hodos 1996; Nieuwenhuys 2002.
271 *What Willis saw:* Dow 1940.
271 *Willis made errors:* Neuburger 1981.
271 *the Scottish physician Robert Whytt:* French 1969.
272 *Staining pieces of brain:* Shepherd 1991.
273 *the damage he did:* Neuburger and Clarke 1981.

273 *To divide the soul:* Clarke and Jacyna 1987, p. 281.
273 *Paul Broca:* Brazier 1988.
274 *Consider a monkey:* Thorpe and Fabre-Thorpe 2001.
275 *a revolutionary technology:* Sowell et al. 1999; Toga and Thompson 2001; Thompson et al. 2001; Beaulieu 2002.
275 *oxygen molecules:* Logothetis et al. 2001.
275 *instead of verbs:* Raichle 1999.
276 *jokes:* Goel and Dolan 2001.
276 *far-flung network:* See, for example Hirsch, Moreno, and Kim 2001.
276 *fueling the brain:* Lennie 2003.
276 *boosts their signal:* Kastner and Ungerleider 2000.
276 *dopamine:* Schultz, Tremblay, and Hollerman 2000.
277 *A gambler shouts for joy:* Wise 2002.
277 *prefrontal cortex:* Botvinick et al. 2001; Miller and Cohen 2001; Miller, Freedman, and Wallis 2002.
279 *Human emotions are descended:* Panskepp 1998.
279 *In both human and mouse:* Panskepp 1998; Insel and Young 2001.
279 *Willis reduced the connection:* Meier 1982.
279 *emotional signals:* LeDoux 2000; Phan et al. 2002.
280 *amgydala:* Amaral 2002.
280 *encodes the primal fears:* Sheline et al. 2001.
280 *a little brain unto itself:* LeDoux 2000.
280 *orbitofrontal cortex:* Montague and Berns 2002.
280 *chocolate:* Small et al. 2001.
280 *money:* O'Doherty et al. 2001.
280 *"why":* Schultz, Tremblay, and Hollerman 2000.
280 *Emotions sharpen our senses:* Dolan 2002.
280 *moderating the emotions:* Beauregard, Levesque, and Bourgouin 2001; Schaefer et al. 2002; Davidson et al. 2002.
281 *upsetting words:* Davidson, Putnam, and Larson 2000.
281 *obsessive-compulsive disorder:* Rosenberg and Macmillan 2002; Saxena, Bota, and Brody 2001.
281 *depressed people:* Elliott et al. 2002.
281 *"shut the little mouths":* Willis 1683, p. 196.
282 *"judicial kindness":* Quoted in Shorter 1997, p. 20.
282 *neurologist named Sigmund Freud:* Shepherd 1991.
283 *an epic of the soul:* Lothane 1998.
283 *Psychoanalysis:* Healy 2002.
283 *They shied away:* Dobson 2001.
283 *"already begun to adopt":* Quoted in Shorter 1997, p. 270.

283 *most psychiatric residents:* Shorter 1997, p. 307.
284 *After six to eight weeks:* Nemeroff and Owens 2002.
284 *scientists are still debating:* See, for example, Duman 2002.
285 *placebo effect:* Kirsch 2002.
285 *William Hogarth:* Hogarth and Shesgreen 1973; MacDonald 1990 discusses it as an emblem of Enlightenment views on madness.
286 *the scans are embraced:* Dumit 1997.
286 *Patients-rights groups:* Valenstein 1998, p. 178.
286 *with depression:* Drevets quoted in Vastag 2002.
287 *"I'm back to being me":* GlaxoSmithKline's drug PaxilCR. See the PaxilCR Web site (http://www.paxilcr.com/). Accessed October 26, 2002.
287 *healthy people:* Knutson et al. 1998; Tse and Bond 2002.
287 *Reasoning:* Shuren and Grafman 2002.
287 *abstract rules:* Miller, Freedman, and Wallis 2002.
288 *This math zone:* Simon et al. 2002; Dehaene 2002.
288 *"consciousness" and "self":* Moutoussis and Zeki 2002.
288 *the semantic mud:* Carter 2002.
288 *synchronization:* Engel and Singer 2001; but see Mazurek and Shadlen 2002.
288 *All animals:* Grush 1997; Damasio 2003.
288 *memories:* Knight and Grabowecky 2000; Tulving 2001.
289 *an awesome social computer:* Lieberman 2000.
289 *a series of unconscious judgments:* Adolphs 2001.
289 *a mental parasite:* Zimmer 2003.
289 *A divining rod:* Wegner 2002.
291 *"Just because!":* Greene 2002.
291 *"immediate feeling":* Quoted in Haidt 2001, p. 3.
291 *"Origin of man now proved":* Barrett et al. 1987, p. 539.
291 *a new theory of moral judgment:* Haidt 2001.
291 *A zookeeper once witnessed this proto-morality:* Waal 1996.
292 *psychopathic criminals:* Blair 1995.
293 *The scans matched his predictions:* Greene et al. 2001.
293 *three other areas:* Greene and Haidt 2002.
295 *anterior cingulate cortex:* Greene et al. 2002.
296 *the placebo effect:* Holden 2002; Leuchter et al. 2002; Mayberg et al. 2002; Petrovic et al. 2002.
296 *psychotherapy and antidepressants:* Brody et al. 2001.

References

Abbott, A. 2002. Neuroscience: Addicted. *Nature* 419 (6910):872–74.

Adami, C. 2002. What is complexity? *Bioessays* 24:1085–94.

Adams, John C. 1999. Sir William Petty: Scientist, economist, inventor, 1623–1687. *Historian* 62:12–15.

Adolphs, R. 2001. The neurobiology of social cognition. *Current Opinions in Neurobiology* 11 (2):231–39.

Albright, T. D., T. M. Jessell, E. R. Kandel, and M. I. Posner. 2000. Neural science: A century of progress and the mysteries that remain. *Cell* 100 Suppl:S1–55.

Alexander, Peter. 2000a. Robert Boyle. In *The dictionary of seventeenth-century British philosophers,* edited by A. Pyle. Bristol, U.K.: Thoemmes Press.

———. 2000b. William Petty. In *The dictionary of seventeenth-century British philosophers,* edited by A. Pyle. Bristol, U.K.: Thoemmes Press.

Allestree, Richard, and John Fell. 1684. *Forty sermons: Whereof twenty one are now first publish'd, the greatest part preach'd before the King and on solemn occasions . . . to these is prefixt an account of the author's life.* Oxford.

Amaral, D. G. 2002. The primate amygdala and the neurobiology of social behavior: Implications for understanding social anxiety. *Biological Psychiatry* 51 (1):11–17.

Anderson, Peter A. V. 1990. *Evolution of the first nervous systems.* New York: Plenum Press.

Anderson, Peter A. V., and Robert M. Greenberg. 2001. Phylogeny of ion channels: Clues to structure and function. *Comparative Biochemistry and Physiology Part B: Biochemistry and Molecular Biology* 129 (1):17–28.

Andrade, E. N. 1950. Wilkins lecture: Robert Hooke. *Proceedings of the Royal Society of London* 137:153–87.

Anonymous. 1672. An account of some books. *Philosophical Transactions of the Royal Society* 7:4071–78.

———. 1683. *Rawleigh Redivivus*. London.

Anstey, Peter. 2000. *The philosophy of Robert Boyle*. London: Routledge.

———. 2001. Boyle against thinking matter. In *Late medieval and early modern corpuscular matter theories,* edited by C. Luthy, J. E. Murdoch, and W. R. Newman. Leiden: E. J. Brill.

Ashley, Maurice. 1990. *The English Civil War.* Gloucester, U.K.: Sutton.

Aspromourgos, Tony. 1996. *On the origins of classical economics: Distribution and value, from William Petty to Adam Smith,* Routledge studies in the history of economics, vol. 6. London: Routledge.

Aubrey, John, and Andrew Clark. 1898. *"Brief lives," chiefly of contemporaries, set down by John Aubrey, between the years 1669 & 1696,* edited by Andrew Clark. Oxford: Clarendon Press.

Bargh, J. A., P. M. Gollwitzer, A. Lee-Chai, K. Barndollar, and R. Trotschel. 2001. The automated will: Nonconscious activation and pursuit of behavioral goals. *Journal of Personality and Social Psychology* 81:1014–27.

Barrett, Paul H., Peter J. Gautrey, Sandra Herbert, David Kohn, and Sydney Smith, eds. 1987. *Charles Darwin's notebooks, 1836–1844: Geology, transmutation of species, metaphysical enquiries.* Ithaca, N.Y.: Cornell University Press.

Bates, Donald G. 1965. Thomas Willis and the epidemic fever of 1661: A commentary. *Bulletin of the History of Medicine* 39:393–414.

Beaulieu, A. 2002. A space for measuring mind and brain: Interdisciplinarity and digital tools in the development of brain mapping and functional imaging, 1980–1990. *Brain and Cognition* 49 (1):13–33.

Beauregard, M., J. Levesque, and P. Bourgouin. 2001. Neural correlates of conscious self-regulation of emotion. *Journal of Neuroscience* 21 (18):RC165.

Bennett, J. A. 1982. *The mathematical science of Christopher Wren.* Cambridge, U.K.: Cambridge University Press.

Blair, R. J. 1995. A cognitive developmental approach to morality: Investigating the psychopath. *Cognition* 57:1–29.

Bloch, Marc. 1973. *The royal touch: Sacred monarchy and scrofula in England and France,* translated by J. E. Anderson. London: Routledge and Kegan Paul.

Bono, James J. 1995. *The word of God and the languages of man: Interpreting nature in early modern science and medicine.* Madison: University of Wisconsin Press.

Botvinick, M. M., T. S. Braver, D. M. Barch, C. S. Carter, and J. D. Cohen. 2001. Conflict monitoring and cognitive control. *Psychological Review* 108 (3):624–52.

Boyle, Marjorie O'Rourke. 2002. Pure of heart: From ancient rites to Renaissance Plato. *Journal of the History of Ideas* 63:41–62.

Boyle, Robert. 1996. *A free enquiry into the vulgarly received notion of nature,* edited by E. B. Davis and M. Hunter. Cambridge, U.K.: Cambridge University Press.

———. 1999. *The works of Robert Boyle,* edited by M. Hunter and E. Davis. London: Pickering & Chatto.

———. 2001. *Correspondence of Robert Boyle, 1636–1691,* edited by M. Hunter, A. Clericuzio, and L. Principe. London: Pickering & Chatto.

Brain, Lord. 1963. The concept of hysteria in the time of William Harvey. *Proceedings of the Royal Society of Medicine* 56:317.

Braithwaite, William C. 1955. *The beginnings of Quakerism,* 2nd ed. Cambridge, U.K.: Cambridge University Press.

Brazier, Mary Agnes Burniston. 1988. *A history of neurophysiology in the 19th century.* New York: Raven Press.

Brody, A. L., S. Saxena, M. A. Mandelkern, L. A. Fairbanks, M. L. Ho, and L. R. Baxter. 2001. Brain metabolic changes associated with symptom factor improvement in major depressive disorder. *Biological Psychiatry* 50 (3):171–78.

Brown, S. P., and R. A. Johnstone. 2001. Cooperation in the dark: Signalling and collective action in quorum-sensing bacteria. *Proceedings of the Royal Society of London. Series B: Biological Sciences* 268 (1470): 961–65.

Brown, Theodore M. 1968. The mechanical philosophy and the "animal oeconomy"—a study in the development of the English physiology in the seventeenth and early eigheenth century. Ph. D. Dissertation, Department of History, Princeton University, Princeton.

———. 1970. The college of physicians and the acceptance of iatromechanism in England, 1665–1695. *Bulletin of the History of Medicine* 44:12–30.

Burnet, Gilbert. 1823. *Bishop Burnet's history of his own time,* edited by M. J. Routh and T. Burnet. Oxford: Clarendon Press.

Burns, Norman T. 1972. *Christian mortalism from Tyndale to Milton.* Cambridge, Mass.: Harvard University Press.

Butler, Ann B., and William Hodos. 1996. *Comparative vertebrate neuroanatomy: Evolution and adaptation.* New York: Wiley-Liss.

Bylebyl, J. J. 1982. Boyle and Harvey on the valves in the veins. *Bulletin of the History of Medicine* 56 (3):351–67.

Bynum, William F. 1973. The anatomical method, natural theology and the functions of the brain. *Isis* 64:445–68.

Canguilhem, Georges. 1993. *A vital rationalist: Selected writings from Georges Canguilhem*, edited by F. Delaporte. New York: Zone Books.

Canny, Nicholas P. 1982. *The upstart earl: A study of the social and mental world of Richard Boyle, first earl of Cork, 1566–1643*. Cambridge: Cambridge University Press.

Capp, B. S. 1994. *The world of John Taylor, the water-poet, 1578–1653*. Oxford: Oxford University Press.

Carew, Thomas. 1949. *The Poems of Thomas Carew,* edited by Rhodes Dunlap. Oxford: Clarendon Press.

Carter, Rita. 2002. *Consciousness*. London: Cassell.

Chen, G. Q., C. Cui, M. L. Mayer, and E. Gouaux. 1999. Functional characterization of a potassium-selective prokaryotic glutamate receptor. *Nature* 402 (6763):817–21.

Christie, W. D. 1871. *A life of Anthony Cooper, first earl of Shaftesbury.* London and New York: Macmillan.

Clark, Stuart. 1997. *Thinking with demons: The idea of witchcraft in early modern Europe*. Oxford: Clarendon Press.

Clarke, Basil. 1975. *Mental disorder in earlier Britain: Exploratory studies.* Cardiff: University of Wales Press.

Clarke, Edwin. 1963. Aristotelian concepts of the form and function of the brain. *Bulletin of the History of Medicine* 37:1–14.

Clarke, Edwin, and L. S. Jacyna. 1987. *Nineteenth-century origins of neuroscientific concepts*. Berkeley: University of California Press.

Clarke, Edwin, and Charles Donald O'Malley. 1996. *The human brain and spinal cord: A historical study illustrated by writings from antiquity to the twentieth century.* San Francisco: Norman Pub.

Clarke, Edwin, and Jerry Stannard. 1963. Aristotle on the anatomy of the brain. *Journal of the History of Medicine* 18:130–48.

Claus, David B. 1981. *Toward the soul: An inquiry into the meaning of psyche before Plato*. New Haven: Yale University Press.

Clericuzio, Antonio. 1993. From van Helmont to Boyle. A study of the transmission of Helmontian chemical and medical theories in seventeenth-century England. *British Journal of the History of Science* 26:303–34.

———. 1994. The internal laboratory. The chemical reinterpretation of medical spirits in England (1650–1680). In *Alchemy and chemistry in the*

16th and 17th centuries, edited by P. Rattansi and A. Clericuzio. Dordrecht, Netherlands: Kluwer.

———. 2000. *Elements, principles and corpuscles: A study of atomism and chemistry in the seventeenth century.* Dordrecht, Netherlands: Kluwer.

Clower, W. T. 1998. The transition from animal spirits to animal electricity: A neuroscience paradigm shift. *Journal of the History of Neuroscience* 7 (3):201–18.

Conway, Anne. 1930. *Conway letters: The correspondence of Anne, viscountess Conway, Henry More, and their friends, 1642–1684,* edited by M. H. Nicolson. New Haven: Yale University Press.

———. 1992. *The Conway letters: The correspondence of Anne, viscountess Conway, Henry More, and their friends, 1642–1684,* edited by M. H. Nicolson and S. Hutton. Rev. with an introduction and new material. Oxford: Oxford University Press.

———. 1996. *The principles of the most ancient and modern philosophy,* edited by A. Coudert. Cambridge texts in the history of philosophy. Cambridge, U.K.: Cambridge University Press.

Cooke, A. M. 1975. William Harvey at Oxford. *Journal of the Royal College of Physicians of London* 9 (2):181–88.

Cowie, Leonard W. 1970. *Plague and fire, London 1665–6.* London: Wayland.

Cranefield, Paul F. 1961. A seventeenth century view of mental deficiency and schizophrenia: Thomas Willis on "stupidity or foolishness." *Bulletin of the History of Medicine* 35:291–316.

Cranston, Maurice W. 1957. *John Locke, a biography.* London: Longmans.

Crombie, A. C. 1967. The mechanistic hypothesis and the scientific study of vision: Some optical ideas as a background to the invention of the microscope. In *Historical aspects of microscopy,* edited by S. Bradbury and G. L. E. Turner. Cambridge, U.K.: Heffer.

Crooke, Helkiah. 1631. *Mikrokosmographia: A description of the body of man: Together with the controversies thereto belonging,* 2nd ed. London.

Cunningham, Andrew. 1997. *The anatomical renaissance: The resurrection of the anatomical projects of the ancients.* Aldershot, U.K.: Ashgate.

Damasio, Antonio R. 2003. Mental self: The person within. *Nature* 423:227.

Davidson, R. J., D. Pizzagalli, J. B. Nitschke, and K. Putnam. 2002. Depression: Perspectives from affective neuroscience. *Annual Review of Psychology* 53:545–74.

Davidson, R. J., K. M. Putnam, and C. L. Larson. 2000. Dysfunction in the neural circuitry of emotion regulation—a possible prelude to violence. *Science* 289 (5479):591–94.

Davis, Audrey B. 1973. *Circulation physiology and medical chemistry in England, 1650–1680*. Lawrence, Kan.: Coronado Press.

Dear, Peter. 1988. *Mersenne and the learning of the schools*. Ithaca, N.Y.: Cornell University Press.

———. 1991. The church and the new philosophy. In *Science, culture, and popular belief in Renaissance Europe,* edited by S. Pumfrey, P. L. Rossi, and M. Slawinski. Manchester, U.K.: Manchester University Press.

———. 2001. *Revolutionizing the sciences: European knowledge and its ambitions, 1500–1700*. Princeton, N.J.: Princeton University Press.

Debus, Allen G. 1970. *Science and education in the seventeenth century: The Webster-Ward debate*. History of science library: Primary sources. London: Macdonald.

———. 1976. *The chemical philosophy: Paracelsian science and medicine in the sixteenth and seventeenth centuries*. New York: Science History Publications.

Dehaene, S. 2002. Single-neuron arithmetic. *Science* 297 (5587):1652–35.

Descartes, René. 1972. *Treatise of man,* translated by T. S. Hall. Harvard monographs in the history of science. Cambridge, Mass.: Harvard University Press.

Dewhurst, Kenneth. 1964. *Thomas Willis as a physician*. William Andrews Clark Memorial Library seminar papers. Los Angeles: William Andrews Clark Memorial Library, University of California.

———. 1966. *Dr. Thomas Sydenham, 1624–1689; his life and original writings*. Berkeley: University of California Press.

———. 1980. *Thomas Willis's Oxford lectures*. Oxford: Sandford Pub.

———. 1984. *John Locke (1632–1704), physician and philosopher*. New York: Garland.

Diamond, Solomon. 1980. Pioneers of psychology: Studies of the great figures who paved the way for the contemporary science of behavior [book review]. *Isis* 71:671–72.

Dobson, Keith S. 2001. *Handbook of cognitive-behavioral therapies*, 2nd ed. New York: Guilford Press.

Doby, Tibor. 1973. Sir Chirstopher Wren and medicine. *Episteme* 4:83–106.

Dolan, R. J. 2002. Emotion, cognition, and behavior. *Science* 298 (5596):1191–94.

Dow, R. 1940. Thomas Willis as a comparative neurologist. *Annals of Medical History* 2:181–94.

Drake, Stillman, N. M. Swerdlow, and Trevor Harvey Levere. 1999. *Essays on Galileo and the history and philosophy of science*. Toronto: University of Toronto Press.

Duman, R. S. 2002. Synaptic plasticity and mood disorders. *Molecular Psychiatry* 7 Suppl 1:S29–34.

Dumit, Joseph. 1997. A digital image of the category of the person: PET scanning and objective self-fashioning. In *Cyborgs & citadels: Anthropological interventions in emerging sciences and technologies,* edited by G. L. Downey and J. Dumit, Santa Fe, N.M.: School of American Research Press.

Eadie, M. J. 2003a. A pathology of the animal spirits—the clinical neurology of Thomas Willis (1621–1675), Part I—Background, and disorders of intrinsically normal animal spirits. *Journal of Clinical Neuroscience* 10:14–29.

———. 2003b. A pathology of the animal spirits—the clinical neurology of Thomas Willis (1621–1675), Part II—Disorders of intrinsically abnormal animal spirits. *Journal of Clinical Neuroscience* 10:146–57.

Eamon, William. 1994. *Science and the secrets of nature: Books of secrets in medieval and early modern culture.* Princeton, N.J.: Princeton University Press.

Edelstein, Ludwig. 1935. The development of Greek anatomy. *Bulletin of the Institute for History of Medicine* 3:235–48.

Elliott, R., J. S. Rubinsztein, B. J. Sahakian, and R. J. Dolan. 2002. The neural basis of mood-congruent processing biases in depression. *Archives of General Psychiatry* 59 (7):597–604.

Elmer, Peter. 1986. *The library of Dr. John Webster: The making of a seventeenth-century radical.* London: Wellcome Institute for the Institute of Medicine.

Empedocles. 2001. *The poem of Empedocles: A text and translation with an introduction,* edited by B. Inwood, rev. ed. Toronto: University of Toronto Press.

Engel, Andreas K., and Wolf Singer. 2001. Temporal binding and the neural correlates of sensory awareness. *Trends in Cognitive Sciences* 5:16–25.

Fasnacht, Ruth. 1954. *A history of the city of Oxford.* Oxford: Basil Blackwell.

Feindel, W. 1962. Thomas Willis (1621–1675)—the founder of neurology. *Canadian Medical Association Journal* 87:289–96.

Finger, Stanley. 2000. *Minds behind the brain: A history of the pioneers and their discoveries.* New York and Oxford: Oxford University Press.

Firth, C. H. 1953. *Oliver Cromwell.* World classics. Oxford: Oxford University Press.

Forrester, J. M. 2002. The marvellous network and the history of enquiry into its function. *Journal of the History of Medicine and Allied Sciences* 57 (2):198–217.

Fox, George. 1659. *A primer for the schollers and doctors of Europe.* London.

Frank, Robert G. 1974. The John Ward diaries: Mirror of seventeenth century science and medicine. *Journal of the History of Medicine and Allied Sciences* 29:147–79.

———. 1979. The image of Harvey in commonwealth and restoration England. In *William Harvey and his age: The professional and social context of the discovery of the circulation,* edited by J. J. Bylebyl. Baltimore: Johns Hopkins University Press.

———. 1980. *Harvey and the Oxford physiologists: Scientific ideas and social interaction.* Berkeley: University of California Press.

———. 1990. Thomas Willis and his circle: Brain and mind in seventeenth-century medicine. In *The Languages of psyche: Mind and body in enlightenment thought,* edited by G. S. Rousseau. Berkeley: University of California Press.

———. 1997. Medicine. In *Seventeenth-century Oxford,* edited by N. Tyacke. Oxford: Clarendon Press.

French, Roger K. 1969. *Robert Whytt, the soul, and medicine.* London: Wellcome Institute of the History of Medicine.

———. 1994. *William Harvey's natural philosophy.* Cambridge, U.K.: Cambridge University Press.

———. 1999. *Dissection and vivisection in the European Renaissance: The history of medicine in context.* Aldershot, U.K.: Ashgate.

Galen. 1968. *Galen on the usefulness of the parts of the body.* Cornell publications in the history of science, translated with introduction by Margaret Tallmadge May. Ithaca, N.Y.: Cornell University Press.

Gaukroger, Stephen. 1995. *Descartes: An intellectual biography.* Oxford: Oxford University Press.

Gert, Bernard. 1996. Hobbes's psychology. In *The Cambridge companion to Hobbes,* edited by T. Sorell. Cambridge, U.K.: Cambridge University Press.

Giglioni, Guido. 1995. Automata compared: Boyle, Leibniz and the debate on the notion of life and mind. *British Journal of the History of Philosophy* 3:249–78.

Goel, V., and R. J. Dolan. 2001. The functional anatomy of humor: Segregating cognitive and affective components. *Nature Neuroscience* 4 (3):237–38.

Goodey, C. F. 1996. The psychopolitics of learning and disability in seventeenth-century thought. In *From idiocy to mental deficiency: Historical perspectives on people with learning disabilities,* edited by D. Wright and A. Digby. London: Routledge.

———. 1999. Politics, nature and necessity: Were Aristotle's slaves feebleminded? *Political Theory* 27:203–24.

———. In press. Intelligence, heredity, and genes: A historical perspective. In *Encyclopedia of the human genome,* edited by D. N. Cooper. London: Macmillan.

———. Submitted. The conception of intellectual disability: False positives in Paracelsus, Plater, and Willis.

Greene, Joshua D. 2002. The terrible, horrible, no good, very bad truth about morality and what to do about it. Ph.D. dissertation, Department of Philosophy, Princeton University, Princeton, N.J.

Greene, Joshua D., and Jonathan Haidt. 2002. How (and where) does moral judgment work? *Trends in Cognitive Sciences* 6:517–23.

Greene, Joshua D., R. Brian Sommerville, Leigh E. Nystrom, John M. Darley, and Jonathan D. Cohen. 2001. An fMRI investigation of emotional engagement in moral judgment. *Science* 293 (5537):2105–8.

———. 2002. Cognitive and affective conflict in moral judgment. Paper read at Cognitive Neuroscience Society Annual Meeting, San Francisco.

Grush, Rick. 1997. The architecture of representation. *Philosophical Psychology* 10:5–23.

Guerlac, Henry. 1954. The poets' nitre. *Isis* 45:243–55.

Gunther, R. T., and Albert Edward Gunther. 1920. *Early science in Oxford.* Oxford: Oxford University Press.

Haidt, J. 2001. The emotional dog and its rational tail: A social intuitionist approach to moral judgment. *Psychological Review* 108 (4):814–34.

Hall, A. Rupert. 1983. *The revolution in science, 1500–1750.* London: Longman.

Hall, Thomas S. 1969. *Ideas of life and matter.* Chicago: University of Chicago Press.

Harrison, Peter. 2002a. Original sin and the problem of knowledge in early modern Europe. *Journal of the History of Ideas* 63:239–59.

———. 2002b. Voluntarism and early modern science. *History of Science* 40:63–89.

Hartlib, Samuel. 1655. *The reformed Common-wealth of bees.* London.

Harvey, William. 1653. *Anatomical exercitations concerning the generation of living creatures: To which are added particular discourses of births, and of conceptions, &c.* London.

———. 1964. *The anatomical lectures of William Harvey. Prelectiones anatomiae universalis. De musculis.* Edited by Gweneth Whitteridge. Edinburgh: Royal College of Physicians.

———. 1981. *Disputations touching the generation of animals,* translated by Gweneth Whitteridge. Oxford: Blackwell Scientific Publications.

———. 1993. *On the motion of the heart and blood in animals,* translated by R. Willis. Buffalo, N.Y.: Prometheus Books.

Hatfield, Gary. 1995. Remaking the science of mind: Psychology of mind. In *Inventing human science: Eighteenth-century domains,* edited by C. Fox, R. Porter, and R. Wokler. Berkeley: University of California Press.

Hawkins, James Michael. 1995. A most excellent antidote: Thomas Willis, the "Ditribae Duae" and the physician's duty. Master's thesis, University of Alberta, Calgary.

Healy, David. 2002. *The creation of psychopharmacology.* Cambridge, Mass.: Harvard University Press.

Henry, John. 1986. Occult qualities and the experimental philosophy: Active principles in pre-Newtonian matter theory. *History of Science* 24:335–81.

———. 1989. The matter of souls: Medical theory and theology in seventeenth-century England. In *The medical revolution of the seventeenth century,* edited by R. French and A. Wear. Cambridge, U.K.: Cambridge University Press.

———. 1991. Doctors and healers: Popular culture and the medical profession. In *Science, culture, and popular belief in Renaissance Europe,* edited by S. Pumfrey, P. L. Rossi, and M. Slawinski. Manchester, U.K.: Manchester University Press.

Hill, Christopher. 1970. *God's Englishman: Oliver Cromwell and the English revolution.* New York: Dial Press.

———. 1972. *The world turned upside down: Radical ideas during the English revolution.* New York: Viking Press.

Hine, William L. 1984. Mersenne: Naturalism and magic. In *Occult and scientific mentalities in the Renaissance,* edited by B. Vickers. Cambridge, U.K.: Cambridge University Press.

Hippocrates. 1923. Regimen. In *Hippocrates.* Cambridge, Mass.: Harvard University Press.

Hirsch, J., D. R. Moreno, and K. H. Kim. 2001. Interconnected large-scale systems for three fundamental cognitive tasks revealed by functional MRI. *Journal of Cognitive Neuroscience* 13 (3):389–405.

Hobbes, Thomas. 1990. *Behemoth; or, The long Parliament,* edited by Ferdinand Tonnies and Stephen Holmes. Chicago: University of Chicago Press.

———. 1994. *The elements of law, natural and politic,* edited by J. C. A. Gaskin.Oxford: Oxford University Press.

Hogarth, William. 1973. *Engravings by Hogarth,* edited by Sean Shesgreen. New York: Dover.

Holcombe, W. M. L., and Ray Paton. 1998. *Information processing in cells and tissues.* New York: Plenum Press.

Holden, Constance. 2002. Drugs and placebos look alike in the brain. *Science* 295:947.

Holland, Guy. 1653. *The grand prerogative of humane nature, namely the soul's natural . . . immortality shewed by many arguments.* London.

Hooke, Robert. 1665. *Micrographia, or, Some physiological descriptions of minute bodies made by magnifying glasses with observations and inquiries thereupon.* London.

Houghton, Walter E. 1942. The English virtuoso in the seventeenth century. *Journal of the History of Ideas* 3:51–73 and 190–219.

Hughes, J. Trevor. 1991. *Thomas Willis 1621–1675: His life and work.* London: Royal Society of Medicine Services.

Hunter, Michael. 1985. The problem of "atheism" in early modern England. *Transactions of the Royal Historical Society* (5th Series) 35:135–57.

———. 1990a. Alchemy, magic, and moralism in the thought of Robert Boyle. *British Journal of the History of Science* 23:387–410.

———. 1990b. Science and heterodoxy: An early modern problem reconsidered. In *Reappraisals of the scientific revolution,* edited by D. C. Lindberg and R. S. Westman. New York: Cambridge University Press.

———. 1995. *Science and the shape of orthodoxy: Intellectual change in late seventeenth-century Britain.* Woodbridge, U.K.: Boydell Press.

———. 1999. Robert Boyle (1627–1691): A suitable case for treatment? *British Journal of the History of Science* 32:261–75.

———. 2000. *Robert Boyle, 1627–91: Scrupulosity and science.* Woodbridge, U.K.: Boydell Press.

Hunter, Richard A., and Ida Macalpine. 1957. Willim Harvey: His neurological and psychiatric observations. *Journal of the History of Medicine and Allied Sciences* 17:126–39.

Hutton, Ronald. 1985. *The Restoration: A political and religious history of England and Wales, 1658–1667.* Oxford: Oxford University Press.

———. 1989. *Charles the second, king of England, Scotland, and Ireland.* Oxford: Oxford University Press.

Hutton, Sarah. 1996. Of physic and philosophy: Anne Conway, F. M. van Helmont and seventeenth-century medicine. In *Religio medici: Medicine and religion in seventeenth century England,* edited by O. P. Grell and A. Cunningham. Aldershot, U.K.: Scolar Press.

———. 1997. Anne Conway, Margaret Cavendish and seventeenth-century scientific thought. In *Women, science and medicine 1500–1700: Mothers and sisters of the Royal Society,* edited by L. Hunter and S. Hutton. Thrupp, U.K.: Sutton.

Hutton, Sarah, and Robert Crocker. 1990. *Henry More (1614–1687).* Tercentenary studies. Dordrecht, Netherlands: Kluwer.

Insel, T. R., and L. J. Young. 2000. Neuropeptides and the evolution of social behavior. *Current Opinions in Neurobiology* 10 (6):784–89.

———. 2001. The neurobiology of attachment. *Nature Reviews Neuroscience* 2 (2):129–36.

Isler, Hansreudi. 1968. *Thomas Willis 1621–75: Doctor and scientist.* New York: Hafner.

———. 1986. Thomas Willis' two chapters on headache of 1672: A first attempt to apply the "new science" to this topic. *Headache* 26:95–98.

Jackson, Stanley W. 1986. *Melancholia and depression: From Hippocratic times to modern times.* New Haven: Yale University Press.

James, Susan. 1997. *Passion and action: The emotions in seventeenth-century philosophy.* Oxford: Oxford University Press.

Jardine, Lisa. 2002. *On a Grander Scale: The Outstanding Life of Sir Christopher Wren.* New York: Harper Collins.

Jobe, T. H. 1976. Medical theories of melancholia in the seventeenth and early eighteenth centuries. *Clio Medica* 11 (4):217–31.

Johns, Adrian. 1996. The physiology of reading and the anatomy of enthusiasm. In *Religio medici: Medicine and religion in seventeenth-century England,* edited by O. P. Grell and A. Cunningham. Aldershot, U.K.: Scolar Press.

Jones, Howard. 1989. *The Epicurean tradition.* London: Routledge.

Jones, Richard Foster. 1961. *Ancients and moderns: A study of the rise of the scientific movement in seventeenth-century England.* St. Louis: Washington University Press.

Kahr, Brett. 1999. Robert Boyle: A Freudian perspective on an eminent scientist. *British Journal of the History of Science* 32:277–84.

Kaplan, Barbara B. 1993. *"Divulging of useful truths in physick": The medical agenda of Robert Boyle.* Baltimore: Johns Hopkins University Press.

Kassler, Jamie. 2001. *Music, science, philosophy: Models in the universe of thought.* Aldershot, U.K.: Ashgate.

Kessler, Eckhard. 1988. The intellective soul. In *The Cambridge history of Renaissance philosophy,* edited by C. B. Schmitt, Q. Skinner, and E. Kessler. Cambridge, U.K.: Cambridge University Press.

Keynes, Geoffrey. 1966. *The Life of William Harvey.* Oxford: Clarendon Press.

Kirsch, Irving. 2002. The emperor's new drugs: An analysis of antidepressant medication data submitted to the U.S. Food and Drug Administration. *Prevention & Treatment* 5: Article 23, posted July 15, 2002. http://journals.apa.org/prevention/volume5/pre0050023a.html/.

Knight, Robert T., and Marcia Grabowecky. 2000. Prefrontal cortex, time, and consciousness. In *The new cognitive neurosciences,* edited by M. S. Gazzaniga and E. Bizzi. Cambridge, Mass.: MIT Press.

Knutson, B., O. M. Wolkowitz, S. W. Cole, T. Chan, E. A. Moore, R. C. Johnson, J. Terpstra, R. A. Turner, and V. I. Reus. 1998. Selective alteration of personality and social behavior by serotonergic intervention. *American Journal of Psychiatry* 155 (3):373–79.

Lapointe, Francois H. 1970. Origin and evolution of the term "psychology." *American Psychologist* 25:640–46.

Larner, Andrew J. 1987. A portrait of Richard Lower. *Endeavour* 11 (4):205–8.

Laver, A. Bryan. 1978. Miracles no wonder! The mesmeric phenomena and organic cures of Valentine Greatrakes. *Journal of the History of Medicine and Allied Sciences* 33:35–46.

LeDoux, Joseph E. 2000. Emotion circuits in the brain. *Annual Review of Neurosciences* 23:155–84.

Lennie, Peter. 2003. The cost of cortical computation. *Neuron* 13:493–97.

Lennox, William G. 1938. Thomas Willis on narcolepsy. *Archives of Neurology and Psychiatry* 41:348–51.

Leuchter, A. F., I. A. Cook, E. A. Witte, M. Morgan, and M. Abrams. 2002. Changes in brain function of depressed subjects during treatment with placebo. *American Journal of Psychiatry* 159 (1):122–29.

Lieberman, M. D. 2000. Intuition: A social cognitive neuroscience approach. *Psychological Bulletin* 126 (1):109–37.

Lindberg, David C. 1992. *The beginnings of Western science: The European scientific tradition in philosophical, religious, and institutional context, 600 B.C. to A.D. 1450.* Chicago: University of Chicago Press.

Locke, John. 1965. *An essay concerning human understanding,* edited by J. W. Yolton. Everyman's library. New York: Dutton.

Loewenstein, Werner R. 1999. *The touchstone of life: Molecular information, cell communication, and the foundations of life.* New York: Oxford University Press.

Logothetis, N. K., J. Pauls, M. Augath, T. Trinath, and A. Oeltermann. 2001. Neurophysiological investigation of the basis of the fMRI signal. *Nature* 412 (6843):150–57.

Longrigg, James. 1988. Anatomy in Alexandria in the third century B.C. *British Journal of the History of Science* 21:455–88.

———. 1993. *Greek rational medicine: Philosophy and medicine from Alcmaeon to the Alexandrians.* London and New York: Routledge.

Lothane, Z. 1998. Freud's 1895 project: From mind to brain and back again. *Annals of the New York Academy of Sciences* 843:43–65.

Lower, Richard. 1932. *Tractatus de Corde,* translated by K. J. Franklin and edited by A. E. Gunther. Vol. 9, *Early Science in Oxford.* Oxford: Oxford University Press.

MacDonald, Michael. 1981. *Mystical bedlam: Madness, anxiety, and healing in seventeenth-century England.* Cambridge monographs on the history of medicine. Cambridge, U.K.: Cambridge University Press.

———. 1990. Insanity and the realities of history in early modern England.

In *Lectures on the history of psychiatry: The Squibb series,* edited by R. M. Murray and T. H. Turner. London: Gaskell.

Macintosh, J. J. 1991. Robert Boyle on Epicurean atheism and atomism. In *Atoms, pneuma, and tranquility: Epicurean and stoic themes in European thought,* edited by M. J. Osler. Cambridge, U.K.: Cambridge University Press.

———. 2002. Robert Boyle. In *Stanford Encyclopedia of Philosophy,* at http://plato.stanford.edu/entries/boyle.

Maddison, Robert E. W. 1969. *The life of the Honourable Robert Boyle.* London: Taylor & Francis.

Martensen, Robert L. 1993. The Circles of Willis: Physiology, culture, and the formation of the "neurocentric" body in England, 1640–1690. Ph.D. dissertation, University of California, San Francisco.

———. 1995. Alienation and the production of strangers: Western medical epistemology and the architectonics of the body. An historical perspective. *Culture, Medicine, and Psychiatry* 19 (2):141–82.

———. 1999. Thomas Willis and the formation of the cerebral body in seventeenth century England. In *A short history of neurology,* edited by F. C. Rose. London: Butterworth-Heineman.

———. 2002. Hippocrates and the politics of medical knowledge in early modern England. In *Reinventing Hippocrates,* edited by D. Cantor. Aldershot, U.K.: Ashgate.

Martinich, Aloysius. 1999. *Hobbes: A biography.* Cambridge, U.K.: Cambridge University Press.

Matthews, Gareth. 2000. Internalist reasoning in Augustine for mind-body dualism. In *Psyche and Soma: Physicians and metaphysicians on the mind-body problem from Antiquity to Enlightenment,* edited by J. P. Wright and P. Potter. Oxford: Oxford University Press.

Mayberg, H. S., J. A. Silva, S. K. Brannan, J. L. Tekell, R. K. Mahurin, S. McGinnis, and P. A. Jerabek. 2002. The functional neuroanatomy of the placebo effect. *American Journal of Psychiatry* 159 (5):728–37.

Mazurek, M. E., and M. N. Shadlen. 2002. Limits to the temporal fidelity of cortical spike rate signals. *Nature Neuroscience* 5 (5):463–71.

McGuire, J. E. 1972. Boyle's conception of nature. *Journal of the History of Ideas* 33:523–42.

Meier, Richard Y. 1979. "Sympathy" as a concept in early neurophysiology. Ph.D. dissertation, Department of History, University of Chicago.

———. 1982. "Sympathy" in the neurophysiology of Thomas Willis. *Clio Medica* 17 (2–3):95–111.

Mendelsohn, J. Andrew. 1992. Alchemy and politics in England. *Past and Present* 133:30–78.

Mercer, Christia. 1993. The vitality and importance of early modern Aristotelianism. In *The rise of modern philosophy: The new and traditional philosophies from Machiavelli to Leibniz,* edited by T. Sorell. Oxford: Oxford University Press.

Merchant, Carolyn. 1980. *The death of nature: Women, ecology, and the scientific revolution.* San Francisco: Harper & Row.

Meyer, A., and R. Hierons. 1965a. On Thomas Willis's concepts of neurophysiology (Part One). *Medical History* 9:1–15.

———. 1965b. On Thomas Willis's concepts of neurophysiology (Part Two). *Medical History* 9:142–55.

Michael, Emily. 2000. Renaissance theories of body, mind, and soul. In *Psyche and soma: Physicians and metaphysicians on the mind-body problem from Antiquity to Enlightenment,* edited by J. P. Wright and P. Potter. Oxford: Oxford University Press.

Michael, Emily, and Fred S. Michael. 1989a. Corporeal ideas in seventeenth-century philosophy. *Journal of the History of Ideas* 50:31–48.

———. 1989b. Two early modern concepts of the mind: Reflecting substance vs. thinking substance. *Journal of the History of Philosophy* 27:29–48.

Miller, E. K., and J. D. Cohen. 2001. An integrative theory of prefrontal cortex function. *Annual Review of Neuroscience* 24:167–202.

Miller, E. K., D. J. Freedman, and J. D. Wallis. 2002. The prefrontal cortex: Categories, concepts and cognition. *Philosophical Transactions of the Royal Society London B* 357 (1424):1123–36.

Milton, J. R. 1994. Locke at Oxford. In *Locke's philosophy: Content and context,* edited by G. A. J. Rogers. Oxford: Oxford University Press.

———. 2001. Locke, medicine and the mechanical philosophy. *British Journal for the History of Philosophy* 9:221–43.

Mintz, Samuel I. 1962. *The hunting of Leviathan: Seventeenth-century reactions to the materialism and moral philosophy of Thomas Hobbes.* Cambridge, U.K.: Cambridge University Press.

Montague, P. R., and G. S. Berns. 2002. Neural economics and the biological substrate of valuations. *Neuron* 36:265–84.

More, Henry. 1653. *An antidote against atheisme, or, An appeal to the natural faculties of the minde of man, whether there be not a God.* London.

More, Louis Trenchard. 1944. *The life and works of the Honourable Robert Boyle.* Oxford: Oxford University Press.

Moutoussis, K., and S. Zeki. 2002. The relationship between cortical activation and perception investigated with invisible stimuli. *Proceedings of the National Academy of Sciences* 99 (14):9527–32.

Nemeroff, C. B., and M. J. Owens. 2002. Treatment of mood disorders. *Nature Neuroscience* 5, Suppl 1:1068–70.

Neuburger, Max. 1981. *The historical development of experimental brain and spinal cord physiology before Flourens.* Baltimore: Johns Hopkins University Press.

Neugebauer, R. 1978. Treatment of the mentally ill in medieval and early modern England: A reappraisal. *Journal of the History of Behavioral Sciences,* edited by Edwin Clarke. 14 (2):158–69.

Newman, William R. 1994. *Gehennical fire: The lives of George Starkey, an American alchemist in the scientific revolution.* Cambridge, Mass.: Harvard University Press.

———. 2001. Experimental corpuscular theory in Aristotelian alchemy: From Geber to Sennert. In *Late medieval and early modern corpuscular matter theories,* edited by C. Luthy, J. E. Murdoch, and W. R. Newman. Leiden: E. J. Brill.

Nieuwenhuys, R. 2002. Deuterostome brains: Synopsis and commentary. *Brain Research Bulletin* 57 (3–4):257–70.

Nutton, Vivian. 2002. Logic, learning, and experimental medicine. *Science* 295:800–801.

O'Connor, J. P. 2003. Thomas Willis and the background to *Cerebri Anatome. Journal of the Royal Society of Medicine* 96:139–43.

O'Doherty, J., M. L. Kringelbach, E. T. Rolls, J. Hornak, and C. Andrews. 2001. Abstract reward and punishment representations in the human orbitofrontal cortex. *Nature Neuroscience* 4 (1):95–102.

Oldenburg, Henry. 1965. *Correspondence,* edited by A. Rupert Hall and Marie Boas Hall. Madison: University of Wisconsin Press.

O'Malley, Charles Donald. 1970. Andreas Vesalius. In *Dictionary of Scientific Biography,* edited by C. C. Gillespie. New York: Scribner.

Osler, Margaret J. 1994. *Divine will and the mechanical philosophy: Gassendi and Descartes on contingency and necessity in the created world.* Cambridge, U.K.: Cambridge University Press.

Oster, Malcolm R. 1989. The "Beam of Diuinity": Animal suffering in the early thought of Robert Boyle. *British Journal of the History of Science* 22:151–80.

Overton, Richard. 1655. *Man wholly mortal, or, A treatise wherein 'tis proved . . . that as whole man sinned, so whole man died.* London.

Pachter, Henry Maximilian. 1951. *Paracelsus: Magic into science.* New York: Schuman.

Pagel, Walter. 1956. Van Helmont's ideas on gastric digestion and gastric acid. *Bulletin of the History of Medicine* 30:524–45.

———. 1970. Johannes (Joan) Baptista van Helmont. In *Dictionary of scientific biography,* edited by C. C. Gillespie. New York: Scribner.

———. 1972. Van Helmont's concept of disease—to be or not to be? The influence of Paracelsus. *Bulletin of the History of Medicine* 46 (5):419–54.

———. 1982. *Joan Baptista van Helmont: Reformer of science and medicine.* Cambridge monographs on the history of medicine. Cambridge, U.K.: Cambridge University Press.

Panskepp, Jaak. 1998. *Affective neuroscience: The foundations of human and animal emotions.* Series in affective science. New York: Oxford University Press.

Parent, Andre. 1986. *Comparative neurobiology of the basal ganglia.* New York: J. Wiley.

Park, Katherine, and Eckhard Kessler. 1988. The concept of psychology. In *The Cambridge history of Renaissance philosophy,* edited by C. B. Schmitt, Q. Skinner, and E. Kessler. Cambridge, U.K.: Cambridge University Press.

Paster, Gail Kern. 1997. Nervous tension: Networks of blood and spirit in the early modern body. In *The body in parts: Fantasies of corporeality in early modern Europe,* edited by D. Hillman and C. Mazzio. New York: Routledge.

Payne, L. M. 1957. Sir Charles Scarburgh's Harveian oration, 1662. *Journal of the History of Medicine* 12:158–64.

Pera, Marcello. 1992. *The ambiguous frog: The Galvani-Volta controversy on animal electricity.* Princeton: Princeton University Press.

Persaud, T. V. N. 1984. *Early history of human anatomy: From antiquity to the beginning of the modern era.* Springfield, Ill.: Thomas.

Peters, R. S. 1956. *Hobbes.* Harmondsworth: Penguin Books.

Petrovic, Predrag, Eija Kalso, Karl Magnus Petersson, and Martin Ingvar. 2002. Placebo and opioid analgesia—imaging a shared neuronal network. *Science* 295:1737–40.

Petty, William. 1927. *The Petty papers; some unpublished writings of Sir William Petty,* edited by H. W. E. P. F. Lansdowne. London: Constable.

Phan, K. L., T. Wager, S. F. Taylor, and I. Liberzon. 2002. Functional neuroanatomy of emotion: A meta-analysis of emotion activation studies in PET and fMRI. *Neuroimage* 16 (2):331–48.

Pizzagalli, D. A., D. Lehmann, A. M. Hendrick, M. Regard, R. D. Pascual-Marqui, and R. J. Davidson. 2002. Affective judgments of faces modulate early activity (approximately 160 ms) within the fusiform gyri. *Neuroimage* 16:663–77.

Popkin, Richard Henry. 1992. *The third force in seventeenth-century thought.* Leiden: E. J. Brill.

Porter, Roy. 1983. The rage of the party: A Glorious Revolution in English psychiatry? *Medical History* 27:35–50.

Porter, S. 1984. The Oxford fire of 1644. *Oxoniensia* 49:289–300.

Principe, Lawrence. 1992. Robert Boyle's alchemical secrecy: Codes, ciphers, and concealments. *Ambix* 39:63–74.

———. 2000. Alchemy, assaying, and experiments. In *Instruments and experimentation in the history of chemistry,* edited by F. L. Holmes and T. H. Levere. Cambridge, Mass.: MIT Press.

Pullman, Bernard. 1998. *The atom in the history of human thought.* Oxford, U.K.: Oxford University Press.

Purves, Dale, and S. Mark Williams. 2001. *Neuroscience.* Sunderland, Mass.: Sinauer Associates.

Raichle, Marcus E. 1998. Behind the scenes of functional brain imaging: A historical and physiological perspective. *Proceedings of the National Academy of Sciences* 95 (3):765–72.

———. 1999. Modern phrenology: Maps of human cortical function. *Annals of the New York Academy of Sciences* 882:107–18; discussion 128–34.

Ramesey, W. 1672. *The gentlemans companion, or, A character of true nobility, and gentility.* London.

Randall, John Herman. 1962. *The career of philosophy: From the Middle Ages to the Enlightenment.* New York: Columbia University Press.

Regev, Aviv, and Ehud Shapiro. 2002. Cells as computation. *Nature* 419:343.

Richards, Graham. 1992. *Mental machinery: The origins and consequences of psychological ideas.* London: Athlone Press.

Rocca, Julius. 1998. Galen and Greek neuroscience. *Early Science and Medicine* 3:216–40.

Rogers, G. A. J. 2000. Thomas Hobbes. In *The dictionary of seventeenth-century British philosophers,* edited by A. Pyle. Bristol, U.K.: Thoemmes Press.

Rosenberg, David R., and Shauna N. Macmillan. 2002. Imaging and neurocircuitry of OCD. In *Neuropsychopharmacology: The fifth generation of progress: An official publication of the American College of Neuropsychopharmacology,* edited by K. L. Davis et al. Philadelphia: Lippincott, Williams & Wilkins.

Ross, Alexander. 1646. *The new planet no planet, or, The earth no wandring star, except in the wandring heads of Galileans.* London.

Rousseau, G. S. 1976. Nerves, spirits, and fibers: Towards defining the origins of sensibility. In *Studies in the Eighteenth Century 3,* edited by R. F. Brissenden and J. C. Eade. Toronto: Toronto University Press.

———. 1991. *Enlightenment crossings: Pre- and post-modern discourses: Anthropological.* Manchester, U.K.: Manchester University Press.

Roy, Ian. 1992. The city of Oxford 1640–1660. In *Town and countryside in the English Revolution,* edited by R. C. Richardson. Manchester, U.K.: Manchester University Press.

Roy, Ian, and Dietrich Reinhart. 1997. Oxford and the civil wars. In *Seventeenth-century Oxford,* edited by N. Tyacke. Oxford, U.K.: Clarendon Press.

Sacks, Oliver W. 1992. *Migraine.* Berkeley: University of California Press.

Sanchez-Gonzalez, M. A. 1990. Medicine in John Locke's philosophy. *Journal of Medicine and Philosophy* 15 (6):675–95.

Sarasohn, Lisa T. 1985. Motion and morality: Pierre Gassendi, Thomas Hobbes, and the mechanical world-view. *Journal of the History of Ideas* 46:363–79.

Sawday, Jonathan. 1995. *The body emblazoned: Dissection and the human body in Renaissance culture.* London and New York: Routledge.

Saxena, S., R. G. Bota, and A. L. Brody. 2001. Brain-behavior relationships in obsessive-compulsive disorder. *Seminars in Clinical Neuropsychiatry* 6 (2):82–101.

Schaefer, S. M., D. C. Jackson, R. J. Davidson, G. K. Aguirre, D. Y. Kimberg, and S. L. Thompson-Schill. 2002. Modulation of amygdalar activity by the conscious regulation of negative emotion. *Journal of Cognitive Neuroscience* 14 (6):913–21.

Schaffer, Simon. 1983. Occultism and Reason. In *Philosophy, its history and historiography,* edited by A. J. Holland. Dordrecht, Netherlands: D. Reidel.

———. 1998. Regeneration: The body of natural philosophers in Restoration England. In *Science incarnate: Historical embodiments of natural knowledge,* edited by C. Lawrence and S. Shapin. Chicago: University of Chicago Press.

Schama, Simon. 2001. *A history of Britain.* New York: Talk Miramax.

Schneider, T. D. 2000. Evolution of biological information. *Nucleic Acids Research* 28:2794–99.

Schultz, W., L. Tremblay, and J. R. Hollerman. 2000. Reward processing in primate orbitofrontal cortex and basal ganglia. *Cerebral Cortex* 10 (3):272–84.

Shapin, Steven. 1991. "A scholar and a gentleman": The problematic identity of scientific practise in early modern England. *History of Science* 29:278–327.

———. 1996. *The scientific revolution.* Chicago: University of Chicago Press.

Shapin, Steven, and Simon Schaffer. 1985. *Leviathan and the air-pump: Hobbes, Boyle, and the experimental life.* Princeton, N.J.: Princeton University Press.

Shapiro, Alan E. 1973. Kinematic optics: A study of the wave theory of light in the seventeenth century. *Archive for History of Exact Sciences* 20:134–270.

Shapiro, Barbara J. 1969. *John Wilkins, 1614–1672: An intellectual biography.* Berkeley: University of California Press.

Shaw, James R. 1972. Models for cardiac structure and function in Aristotle. *Journal of the History of Biology* 5:355–88.

Shea, William R. 1986. Galileo and the church. In *God and nature: Historical essays on the encounter between Christianity and science*, edited by D. C. Lindberg and R. L. Numbers. Berkeley: University of California Press.

Sheline, Y. I., D. M. Barch, J. M. Donnelly, J. M. Ollinger, A. Z. Snyder, and M. A. Mintun. 2001. Increased amygdala response to masked emotional faces in depressed subjects resolves with antidepressant treatment: An fMRI study. *Biological Psychiatry* 50 (9):651–58.

Shepherd, Gordon M. 1991. *Foundations of the neuron doctrine.* New York: Oxford University Press.

Sherrer, Grace B. 1958. Philalgia in Warwickshire: F. M. Van Helmont's anatomy of pain applied to Lady Anne Conway. *Studies in the Renaissance* 5:196–206.

Sherrington, Charles Scott. 1941. *Man on his nature.* New York: Macmillan.

Shorter, Edward. 1997. *A history of psychiatry: From the era of the asylum to the age of Prozac.* New York: John Wiley & Sons.

Shuren, J. E., and J. Grafman. 2002. The neurology of reasoning. *Archives of Neurology* 59 (6):916–19.

Simon, O., J. F. Mangin, L. Cohen, D. Le Bihan, and S. Dehaene. 2002. Topographical layout of hand, eye, calculation, and language-related areas in the human parietal lobe. *Neuron* 33 (3):475–87.

Simpson, William. 1659. *From one who was moved of the Lord God.* London.

Skinner, Quentin. 1966. The ideological context of Hobbes's political thought. *Historical Journal* 9:286–317.

———. 1969. Thomas Hobbes and the nature of the early Royal Society. *Historical Journal* 12:217–39.

Skwire, Sarah E. 1999. Women, writers, sufferers: Anne Conway and Ann Collins. *Literature and Medicine* 18:1–23.

Small, D. M., R. J. Zatorre, A. Dagher, A. C. Evans, and M. Jones-Gotman. 2001. Changes in brain activity related to eating chocolate: From pleasure to aversion. *Brain* 124 (Pt. 9):1720–33.

Smith, C. U. 1998. Descartes' pineal neuropsychology. *Brain and Cognition* 36 (1):57–72.

Sowell, E. R., P. M. Thompson, C. J. Holmes, R. Batth, T. L. Jernigan, and A. W. Toga. 1999. Localizing age-related changes in brain structure between childhood and adolescence using statistical parametric mapping. *Neuroimage* 9 (No. 6, Pt. 1):587–97.

Stephens, Philip. 1658. *Catalogus horti botanici Oxoniensis*. Oxford.

Stevens, Scott M. 1997. Sacred heart and secular brain. In *The body in parts: Fantasies of corporeality in early modern Europe*, edited by D. Hillman and C. Mazzio. New York: Routledge.

Stevenson, Lloyd G. 1965. "New diseases" in the seventeenth century. *Bulletin of the History of Medicine* 39:1–21.

Stillingfleet, Edward. 1667. *A sermon preached before the King, March 13. 1666/7*. London.

Strauss, Erich. 1954. *Sir William Petty: Portrait of a genius*. Glencoe, Ill.: Free Press.

Sutton, John. 1998. *Philosophy and memory traces: From Descartes to connectionism*. Cambridge, U.K.: Cambridge University Press.

Temkin, Owsei. 1971. *The falling sickness: A history of epilepsy from the Greeks to the beginning of modern neurology*, 2nd ed. Baltimore: Johns Hopkins University Press.

———. 1973. *Galenism: Rise and decline of a medical philosophy*. Cornell publications in the history of science. Ithaca, N.Y.: Cornell University Press.

Thomas, Keith. 1971. *Religion and the decline of magic: Studies in popular beliefs in sixteenth and seventeenth century England*. London: Weidenfeld & Nicolson.

Thomas, Peter William. 1969. *Sir John Berkenhead, 1617–1679: A Royalist career in politics and polemics*. Oxford: Oxford University Press.

Thompson, P. M., C. Vidal, J. N. Giedd, P. Gochman, J. Blumenthal, R. Nicolson, A. W. Toga, and J. L. Rapoport. 2001. Mapping adolescent brain change reveals dynamic wave of accelerated gray matter loss in very early–onset schizophrenia. *Proceedings of the National Academy of Sciences* 98 (20):11650–55.

Thorpe, S. J., and M. Fabre-Thorpe. 2001. Seeking categories in the brain. *Science* 291 (5502):260–63.

Tieleman, Teun. 1996. *Galen and Chrysippus on the soul: Argument and refutation in the De Placitis, Books II–III*. Leiden and New York: E. J. Brill.

Tinniswood, Adrian. 2001. *His invention so fertile: A life of Christopher Wren*. London: Jonathan Cape.

Toga, Arthur W., and Paul M. Thompson. 2001. Maps of the brain. *The Anatomical Record* 265:37–53.

Tourney, Garfield. 1972. The physician and witchcraft in Restoration England. *Medical History* 16:143–55.

Trevor-Roper, H. R. 1985. The Paracelsian movement. In *Renaissance essays*. London: Secker & Warburg.

———. 1998. Paracelsianism made political 1600–1650. In *Paracelsus: The man and his reputation, his ideas and their transformation,* edited by O. P. Grell. Leiden: E. J. Brill.

Tse, W. S., and A. J. Bond. 2002. Serotonergic intervention affects both social dominance and affiliative behaviour. *Psychopharmacology* 161 (3):324–30.

Tuck, Richard. 1992. The "Christian Atheism" of Thomas Hobbes. In *Atheism from the Reformation to the Enlightenment,* edited by M. Hunter and D. Wootton. Oxford: Oxford University Press.

Tulving, E. 2001. Episodic memory and common sense: How far apart? *Philosophical Transactions of the Royal Society B Biological Sciences* 356 (1413):1505–15.

Tyacke, Nicholas. 1978. Science and religion at Oxford before the Civil War. In *Puritans and revolutionaries: Essays in seventeenth-century history presented to Christopher Hill,* edited by D. Pennington and K. Thomas. Oxford: Clarendon Press.

———. 1997. Religious controversy. In *Seventeenth-century Oxford,* edited by N. Tyacke. Oxford: Clarendon Press.

Valenstein, Elliot S. 1998. *Blaming the brain: The truth about drugs and mental health.* New York: Free Press.

van der Eijk, P. J. 2000. Aristotle on the soul-body relationship. In *Psyche and soma: Physicians and metaphysicians on the mind-body problem from Antiquity to Enlightenment,* edited by J. P. Wright and P. Potter. Oxford: Oxford University Press.

Vastag, B. 2002. Decade of work shows depression is physical. *Journal of the American Medical Association* 287 (14):1787–88.

Veith, Ilza. 1965. *Hysteria: The history of a disease.* Chicago: University of Chicago.

Vesling, Johann, and Gerardus Leonardus Blasius. 1666. *Syntagma anatomicum.* Amsterdam.

Vickers, Brian. 1984. Introduction. In *Occult and scientific mentalities in the Renaissance,* edited by B. Vickers. Cambridge, U.K.: Cambridge University Press.

Von Staden, Heinrich. 1989. *Herophilus: The art of medicine in early Alexandria.* Cambridge, U.K.: Cambridge University Press.

Vrooman, Jack Rochford. 1970. *René Descartes: A biography.* New York: Putnam.

Waal, F. B. M. de. 1996. *Good natured: The origins of right and wrong in humans and other animals.* Cambridge, Mass.: Harvard University Press.

Walker, D. P. 1985. *Music, spirit and language in the Renaissance.* London: Variorum Reprints.

———. 2000. *Spiritual and demonic magic from Ficino to Campanella*. University Park: Pennsylvania State University Press.

Wallace, Wes. 2003. The vibrating nerve impulse in Newton, Willis and Gassendi: First steps in a mechanical theory of communication. *Brain and Cognition* 51:66–94.

Wallis, John. 1666. A relation of an accident by thunder and lightning, at Oxford. *Philosophical Transactions on the Royal Society of London* 1:222–26.

Watkins, Richard. 1651. *Newes from the dead. Or a true and exact narration of the miraculous deliverance of Anne Greene*. London.

Wear, A. 2000. *Knowledge and practice in English medicine, 1550–1680*. Cambridge, U.K.: Cambridge University Press.

Webster, Charles. 1969. Henry More and Descartes: Some new sources. *British Journal of the History of Science* 4:359–77.

———. 1971. The Helmontian George Thomson and William Harvey: The revival and application of splenectomy to physiological research. *Medical History* 15 (2):154–67.

———. 1973. William Dell and the idea of university. In *Changing perspectives in the history of science: Essays in honour of Joseph Needham*, edited by M. Teich and R. M. Young. London: Heinemann.

———. 1975. *The great instauration: Science, medicine, and reform 1626–1660*. London: Duckworth.

———. 1982. *From Paracelsus to Newton: Magic and the making of modern science*. Cambridge, U.K.: Cambridge University Press.

———. 2002. Paracelsus, Paracelsianism, and the secularization of the worldview. *Science in Context* 15 (1):9–27.

Weeber, Edwin J., and J. David Sweatt. 2002. Molecular neurobiology of human cognition. *Neuron* 33:845–48.

Wegner, Daniel M. 2002. *The illusion of conscious will*. Cambridge, Mass.: MIT Press.

Wilkins, John. 1638. *The discovery of a world in the moone, or, A discourse tending to prove that 'tis probable there may be another habitable world in that planet*. London.

Willis, Thomas. 1664. *Cerebri anatome: Cui accessit nervorum descriptio et usus*. London.

———. 1681a. *An essay of the pathology of the brain and nervous stock in which convulsive diseases are treated of*, translated by Samuel Pordage. London: Printed by J. B. for T. Dring.

———. 1681b. *A medical-philosophical discourse of fermentation, or, Of the intestine motion of particles in every body*, translated by Samuel Pordage. London.

———. 1681c. *The remaining medical works of that famous and renowned physician Dr. Thomas Willis, The first part, though last published,* translated by Samuel Pordage. London.

———. 1683. *Two discourses concerning the soul of brutes, which is that of the vital and sensitive of man. The first is physiological . . . The other is pathological.* Translated by Samuel Pordage. London.

———. 1684. *Pharmaceutice rationalis, or, An exercitation of the operations of medicines in humane bodies shewing the signs, causes, and cures of most distempers incident thereunto: In two parts: As also a treatise of the scurvy, and the several sorts thereof, with their symptoms, causes, and cure,* translated by Samuel Pordage. London.

———. 1691. *A plain and easie method for preserving (by God's blessing) those that are well from the infection of the plague, or any contagious distemper in city, camp, fleet, &c and for curing such as are infected with it: Written in the year 1666.* London.

———. 1965. *The anatomy of the brain and nerves,* edited by W. Feindel. Montreal: McGill University Press.

———. 1981. *Willis's Oxford casebook (1650–52),* edited by K. Dewhurst. Oxford: Sandford Publications.

Wilson, Catherine. 1988. Visual surface and visual symbol: The microscope and the occult in early modern science. *Journal of the History of Ideas* 49:54–108.

———. 2000. Descartes and the corporeal mind: Some implications of the Regius affair. In *Descartes' natural philosophy,* edited by S. Gaukroger, J. A. Schuster, and J. Sutton. London: Routledge.

Wilson, Leonard G. 1993. Fever. In *Companion encyclopedia of the history of medicine,* edited by W. F. Bynum and R. Porter. London: Routledge.

Wise, R. 2002. Brain reward circuitry: Insights from unsensed incentives. *Neuron* 36 (2):229.

Wood, Anthony á. 1891. *The life and times of Anthony Wood, antiquary, of Oxford, 1632–1695.* Oxford: Clarendon Press.

Wright, John P. 1980. Hysteria and mechanical man. *Journal of the History of Ideas* 41:233–47.

———. 1991. Locke, Willis, and the seventeenth-century Epicurean soul. In *Atoms, pneuma, and tranquility: Epicurean and Stoic themes in European thought,* edited by M. J. Osler. Cambridge, U.K.: Cambridge University Press.

Young, Robert M. 1970. *Mind, brain and adaptation in the nineteenth century: Cerebral localization and its biological context from Gall to Ferrier.* Oxford: Clarendon Press.

Zimmer, C. 2003. "How the mind reads other minds." *Science,* 300:1079–80.

Acknowledgments

More than any book I have written before, this one was built on reading. I would therefore like to thank the staff of the libraries where I did much of my research for this book: the Bodleian Library of Oxford University, the British Library, the Wellcome Library for the History and Understanding of Medicine in London, the New York Public Library, the Bobst Library of New York University, and the New York Academy of Medicine Library. I am grateful as well to the scientists who helped me learn about the current state of the neurosciences, particularly Jon Cohen, James Goldman, Joshua Greene, Joy Hirsch, Jean-Paul von Sattel, and Nicholas Schiff. Several experts were kind enough to look over my manuscript, including Peter Anstey, William Feindel, James Goldman, Joshua Greene, Robert Hatch, John Henry, and John R. Milton. My gratitude goes to them, although the responsibility for errors remains with me. I also would like to thank the

John S. Guggenheim Foundation for its generous fellowship, which helped make this book possible.

I thank all my friends who helped in big and small ways. My gratitude goes out to John Zimmer (no relation) and James Stewart for their boundless hospitality on my research visits to England. I am fortunate to have Eric Simonoff as an agent—someone with both a taste for Stuart England and a sharp eye for wandering story lines. I thank both my editors. Stephen Morrow possesses that rare sort of brain that can handle two channels of information at once: the enthusiasm all authors want to hear and the unsparing criticism they all need. Leslie Meredith took on the editing of this book like an equestrian leaping onto a galloping horse and rode it safely home. Thanks also to Ravi Mirchandani, my British editor, for a helping hand that stretched across the ocean.

But most of all, I am fortunate to be married to my wife, Grace, who refashioned our lives so that I could write this book.

Index

About the Author

The New York Times Book Review calls Carl Zimmer "as fine a science essayist as we have." His work appears regularly in *The New York Times, National Geographic, Newsweek, Discover, Natural History,* and *Science.* A John S. Guggenheim Fellow, he has also received the Pan-American Health Organization Award for Excellence in International Health Reporting and the American Institute of Biological Sciences Media Award. His previous books include *Evolution: The Triumph of an Idea; Parasite Rex;* and *At the Water's Edge.* He lives in Guilford, Connecticut.

www.carlzimmer.com